CELLULOSICS UTILIZATION

Research and Rewards in Cellulosics

Proceedings of the Nisshinbo International Conference on Cellulosics Utilization in the Near Future, 5–6 December 1988, Tokyo, Japan (a Post-Symposium of Cellucon '88 Japan), organized by Nisshinbo Industries, Inc.

ORGANIZING COMMITTEE

General Chairman: Inagaki, Hiroshi (Kyoto Univ.)
Vice General Chairman: Sohma, Isamu (Nisshinbo Ind.)
Chairman: Hongu, Tatsuya (Nisshinbo Ind.)
Members: Komiyama, Jiro (Tokyo Inst. Tech.)
Komoto, Tadashi (Gunma Univ.)
Miyamoto, Takeaki (Kyoto Univ.)
Nakamura, Yoshio (Gunma Univ.)
Onabe, Fumihiko (Univ. Tokyo)
Phillips, Glyn O. (N.E. Wales Inst.)
Tanioka, Akihiko (Tokyo Inst. Tech.)
Yamada, Akira (Inst. Phys. Chem. Res.)
Kimura, Ken (Nisshinbo Ind.)
Suzuki, Toshio (Nisshinbo Ind.)
Shimizu, Toshio (Nisshinbo Ind.)
Shimada, Shoichi (Nisshinbo Ind.)
Yanai, Yuichi (Nisshinbo Ind.)

CELLULOSICS UTILIZATION

Research and Rewards in Cellulosics

Edited by

H. INAGAKI

Kyoto University, Kyoto, Japan

and

G. O. PHILLIPS

The North East Wales Institute, Clwyd, Wales, UK

ELSEVIER APPLIED SCIENCE
LONDON and NEW YORK

ELSEVIER SCIENCE PUBLISHERS LTD
Crown House, Linton Road, Barking, Essex IG11 8JU, England

Sole Distrubutor in the USA and Canada
ELSEVIER SCIENCE PUBLISHING CO., INC.
655 Avenue of the Americas, New York, NY 10010, USA

WITH 28 TABLES AND 91 ILLUSTRATIONS

© 1989 ELSEVIER SCIENCE PUBLISHERS LTD

British Library Cataloguing in Publication Data

Nisshinbo International Conference on Cellulosics
  Utilization in the Near Future (1988) Tokyo, Japan
  Cellulosics utilization.
  1. Cellulosics
  I. Title II. Inagaki, H. (Hiroshi) III. Phillips,
  Glyn O. (Glyn Owain) *1927–*
  668.4′4

  ISBN 1-85166-406-8

Library of Congress CIP data applied for

Printed in Northern Ireland by The Universities Press (Belfast) Ltd.

FOREWORD

The first encounter of the Japanese with cellulosic products took place in 1877, when commercial samples of celluloid were imported for exhibition from Germany. This led to the production of synthetic coral beads, combs and other ornaments required by the public to replace 'bekko' (tortoise-shell). The domestic production of nitrocellulose did not take place until 1908, when two manufacturers were founded, which later merged into one company (the origin of the present Daicel Chemical Industries, Ltd).

Meanwhile, a pilot plant was built in 1905 to produce viscose to be used as a coating material for synthetic leather to replace the nitrocellulose used conventionally in those days. A sudden rise in the price of imported rayon, due to World War I, accelerated research into viscose to produce rayon in Japan. As a result, the present Teijin Ltd was founded in 1918. Cuprammonium rayon had attracted a large interest from cellulose chemical industries since 1916, but the first commercial samples appeared on the market in 1931, when a large-scale production process was established by the present Asahi Chemical Industries Co., Ltd.

As mentioned above, the basic principles employed in these processes were discovered or invented in Europe or North America, and the Japanese only applied established principles to rationalize a production system by improving quality and developing new applications. In fact, there was no significant cellulose-related research before 1920 in Japan, except for the special contribution made in 1913 by Shoji Nishikawa (1884–1952) who discovered, using X-ray diffractometry, that cellulosic fibrous substances, such as wood, bamboo and hemp, have a crystalline structure.

As the cellulose chemical industry grew to lead domestic industries, the 'Sen'iso Kyokai' (Research Association for Cellulose) was founded in 1925, and its journal Sen'iso Kōgyo (Cellulose Industry) was issued in the same year. This journal was the first academic journal dedicated to cellulose research in Japan. The Japanese cellulose chemical industry was expanded to international standards in terms of cellulosics production in the 1930s. On this occasion, the 'Sen'i Kōgyo Gakkai' (Society of Fiber Industry) was founded in 1935, and another journal was published for fiber technology. The outbreak of the Pacific War in December 1941 caused a serious shortage of commodities. These two institutions were consequently united into the present 'Sen'i Gakkai' (Society of Fiber Science and Technology, Japan) on 10 December 1943, and its journal, the *Journal of the Society of Fiber Science and Technology*, emerged.

The original purpose of the Society was to promote the development of the cellulose chemical industry. However, there was no opportunity to organize any international meetings on cellulose for over half a century. In 1983, on the occasion of the 40th anniversary of our Society, various events were planned to promote the textile industry in Japan. One of them was the 'International Symposium on Fiber Science and Technology' (ISF '85) held at Hakone in August 1985. During the course of the program preparation, three main subjects including one on 'cellulosics' were proposed. This subject 'cellulosics' was chosen to provide an opportunity to recall the history of our Society, as well as to re-evaluate cellulose as a renewable resource and the potential to develop into new fields. The Organizing Committee was not sure at first whether sufficient papers would be submitted for the sessions allocated to 'cellulosics', but eventually the Committee had to cope with 85 applications from inland and also from abroad, including several invited lectures. Such an enthusiastic response stimulated our Society to organize 'Cellucon 88 Japan' in cooperation with Cellucon Conferences.

When our Society decided to hold 'Cellucon 88 Japan' and started its preparation, Mr Tatsuo Tanabe, the President of Nisshinbo Industries, Inc., asked for our cooperation in organizing a post-symposium to celebrate the 80th anniversary of his company. Mr Tanabe's intention was to find a breakthrough for his company into the cellulosics field by making the best use of its long experience in cotton yarn and cotton fabrics production, and to contribute to the advancement of closer international cooperation in cellulose research.

Since then, Professor T. Miyamoto (Kyoto University), Dr T. Hongu (Nisshinbo Industries) and I considered at some length how we should organize such a conference! Two standpoints emerged from our discussion. The first was that the image which the word 'cellulose' usually affords to people should be radically changed. To this end, the Organizing Committee formed for this symposium decided to have two 'Round Table' sessions, which were designated 'Hi-Tech' and 'Bio-Tech', respectively. The second standpoint was that the conference should not confine itself to cellulose but extend to other polysaccharides. By doing so, we expected to amalgamate knowledge from more than one area so that some unique development would emerge.

This book is a collection of lectures read at the International Conference held in Tokyo on 5 and 6 December 1988. Discussions after each lecture were recorded in order to transmit the atmosphere of each session to readers. Finally, I hope that the reader will find out something new and important about cellulosics from this book.

HIROSHI INAGAKI
Professor Emeritus, Kyoto University
Professor of Polymer Chemistry
at Mukogawa Women's University

PREFACE

This book is certain to be valuable to students and researchers in cellulose and polysaccharide chemistry, because the information presented is new and not available collectively in any other publication.

Nisshinbo Industries must be congratulated on their imagination in supporting this International Conference, and drawing together leading polysaccharide chemists from all over the world. The Conference which this book records marks 80 successful years of cellulosics application by the company, and provides realistic signposts for the future. The message of this volume is that cellulose has cast away its middle-aged image: the headings 'Hi-Tech' and 'Bio-Tech' for the Round Table discussions at the meeting, capture the new mood. It is the second time during my own scientific career that cellulose has hit back and surprised its competitors. And I can honestly say that Japan is leading the way in this revival.

Perhaps I may be permitted here a personal note. I was introduced to cellulose at first hand at what was then the USDA Southern Regional Laboratory in New Orleans by my friend Jett C. Arthur Jr in the early 1960s. Then the new synthetic fibres were making inroads and it appeared as though cotton was on the way out. Then, due to the work of that laboratory and the Shirley Institute in Manchester, as much as anywhere, the new chemically modified cottons and blends made a comeback. 'King Cotton Rides Again', proclaimed the *Times Picayune* of New Orleans!

I believe that we are now witnessing a second major breakthrough. We now recognize that we do not have to settle for and utilize the cellulose molecule, from whatever source we find it, in the form we find it. Let me illustrate my claim.

Structure and Functionality

New information is available about what we used to call the 'crystalline' and 'amorphous' regions. There is a beauty, symmetry and distinctiveness in the hydrogen-bonded structures which we code-name Cellulose I and Cellulose II, as shown in the illustration. Knowledge of structure has opened up new vistas. An understanding of the role of bound water and H-bonding, degree of crystallinity, structure of the fibrils, voids, pores and accessibility in controlling reactivity is opening up new practical routes. Enzymic degradation, for example, increases with degree of crystallinity. Kamide and his co-workers have calculated, for example, the positions of lines in the ^{13}C-NMR spectrum

STRUCTURE OF CELLULOSE I :
(a) ac projection ; (b) ab projection

STRUCTURE OF CELLULOSE II:
(a) ac projection ; (b) ab projection

(a)

(a)

(b)

(b)

for different hydrogen bonds. The reactivity of pulps for the viscose process increases with increasing content of $0(2)H...0(6)$ and decreasing content of $0(3)H...0(5)$ intramolecular hydrogen bonds in the cellulose. The drawing of fibres from an aqueous alkali cellulose solution and then converting the fibre into fabrics, gives to wood pulp cellulose the physical benefits of cotton and the manufacturing flexibility of a synthetic polymer. All due to gaining an understanding of the H-bonded systems. Could there be a better example where faith in basic research has led to practical rewards?

Unlocking the Wood Matrix

Wood is, and has been, a constant 'new material': new resources are generated with no help from the chemist and without high technology or environmental contamination. Chemical unlocking of the matrix, however, is not without its difficulties as the pulping chemists amongst us know. I have been fascinated by the wood explosion process; simply introduce steam into wood chips under pressure and the lignin, hemicellulose and cellulose fly apart. Lignin is available for chemical and biological exploitation; the hemicellulose is hydrolysed and microcrystalline cellulose or higher molecular weight cellulose is produced. Could it be the missing key to unlock the stubborn wood matrix? New organic solvent processes (organo-solv) are also making the component materials available in quantity for further chemical exploitation.

New Chemical and Pharmaceutical Products

There is reason now to consider cellulose as a functional chemical, not a bulk commodity material. This book describes a range of new products which show a specific functionality:

- Carboxyethyl carbamoylethyl cellulose to accelerate the clotting of blood.
- Sulphated celluloses to prevent the clotting of blood.
- High absorbency celluloses, which are elastic only when wet.
- High viscosity pseudoplastic hydrophobically modified celluloses.
- Cellulosic liquid crystals.
- Cellulose derivatives to separate racemic compounds.
- Antithrombogenic cellulosic membranes.
- Cellulosic membranes for haemodialysis and haemofiltration.
- Cellulosic sequestering agents.

The book opens the door to a Pandora's box of new products derived from cellulosics and other naturally occurring polysaccharides. It proved a great benefit that the Conference did not confine itself to cellulose. In the past, carbohydrate chemists have been too isolated from the wider polymer and biopolymer fields. As shown in the book, starch, chitin, hyaluronic acid, lentinan and semi-synthetic glucans are now finding widespread new applications in physically and chemically modified forms. With my two friends, Professor Endre A. Balazs of Biomatrix Inc. (USA) and Professor Y. Nakamura at the University of Gunma, Japan, and their colleagues, we have grafted hyaluronic acid on to biological membranes, so conferring on them excellent biocompatibility whilst retaining the strength needed by surgical tissue grafts. Cellulose–synthetic polymer graft co-polymers can now be tailor-made with precise control over their molecular weight, degree of substitution and backbone graft linkage. Biotechnology too is helping to synthesize new celluloses, and modify existing cellulosics for new applications.

The book adequately demonstrates that the imaginative new developments are emerging by integrating knowledge from a variety of disciplines and materials. The reader regretfully will not be able to enjoy the outstanding hospitality and fellowship which characterized the Conference. All our thanks are due to Nisshinbo Industries, and with this book may we wish them a happy 80th birthday and rejoice that, as the company gets older, it is also becoming more innovative and internationally active. *Domo arigato*, or as we say in Wales, *Diolch yn Fawr*.

GLYN O. PHILLIPS
The North East Wales Institute,
Clwyd, Wales, UK

CONTENTS

Special Contributions

<u>PLENARY SESSION</u>

PERSONAL REMINISCENCES AND PERSPECTIVES ON CELLULOSE AND WOOD CHEMISTRY

Conrad Schuerch
Department of Chemistry
SUNY College of Environmental Science and Forestry
Syracuse, New York 13210

ABSTRACT

An informal account of some researches in cellulose, and wood chemistry by the author, his associates and colleagues, their somewhat tortuous ramifications and their relation to present day practice. The general conclusion that may be drawn is that ideas produce results according to the law of unexpected consequences.

PLENARY LECTURE

I should like first to congratulate Nisshinbo Industries on its 80th Anniversary and to thank them, and specifically the organizing committee, for inviting me again to Japan, to speak at this Conference. I am sure you are aware that we were looking forward to a lecture from Professor J. J. Lindberg and we all regret that, due to illness, he can not be with us today. Surely no one regrets his absence more than I do! I have incidentally, an especially warm feeling for Professor Lindberg, since he referred extensively in his doctoral thesis and in his publications on lignin association phenomena to a 1952 publication of mine on solvent properties of liquids and their relation to lignin solubility, swelling and isolation [1]. That was my first major independent contribution and I hold an especial affection for it.

When I was asked to give a plenary lecture at this Conference, I admit it appeared to be a formidable undertaking. I surmised that most of the audience would have already listened for nearly a week to new technical results from the most active research laboratories worldwide and would be looking forward to two days more of exciting research that has great promise of industrial application. That is certainly the case today, so I thought that perhaps a short interlude of personal reminiscences might be a respite and relaxation, and might give a little perspective on how far we have come, and where we may go. Furthermore, those who do not know history are compelled to repeat it, and nothing is more useless than obtaining old research results with new more elegant techniques.

My first experience with conferences in wood chemistry was in 1949. I can remember the date because I had just returned to Montreal from my honeymoon, dropped off my new wife, and left for the First Lignin Round Table in Appleton, Wisconsin at the Institute of Paper Chemistry. The conference was well attended with the foremost lignin chemists of the time. The acknowledged King, Professor Karl Freudenberg, the emerging Crown Prince, Professor Eric Adler, the brilliant Prime Minister Professor Holger Erdtman were all there, I believe. The atmosphere was enthusiastic and filled with the idealism of organic chemists: As soon as we know the structure of lignin, all sorts of wonderful products and applications will follow. Finally on the last day of the conference, a practical paper man who could stand this emphasis no longer, stood up and said, "I'll tell you what to do with lignin. Leave it in the pulp!"

If one looks at the direction the pulp and paper industry has followed since that time, it is clear that that was very sensible advice. There has been a continuing increase in the use of high yield pulps with mechanical, thermomechanical and the various semichemical processes in more forms and more applications. There are good reasons for this development, both practical and scientific. The industry is set up to produce and market paper products, not fine chemicals. The component with highest heat content in pulping liquors is lignin and has a significant fuel value for chemical recovery and so on. But more fundamentally we now know a great deal about the structure of lignin. It is a network polymer based on a number of different monomeric units linked by both carbon to carbon and carbon to oxygen bonds. There is no way to degrade a material like this to anything but a very complex mixture of products, many with intractable or unexceptional properties. It is not surprising, therefore, that with few exceptions, successful lignin applications have all been applications that dealt with lignin as a polymeric material, not a source of industrial or fine chemicals. Nevertheless, one still finds some research that seems to be based on the premise that lignin is a likely source of industrial or fine chemicals. I wish these enthusiastic researchers would read the review of Herrick and Hergert [2], and the contributions of Drs. Pearl, Hearon and many others, and perhaps direct their efforts more profitably elsewhere. What we really need is a magician who can convert lignin in pulp into a more hydrophilic polymer that does not absorb ultraviolet light nor act as a phototendering agent. The appearance of such a magician does not appear to be likely in the near future.

Perhaps I should give you some idea of the research atmosphere at the time I began my career at the College of Forestry in Syracuse in 1949. We were not in bad shape instrumentally for organic chemistry of that day, for we had a Beckman DU spectrophotometer, and a manual polarimeter. However our laboratories in the basement of Bray Hall were old and obsolete. When the Chemistry Department was organized in 1952 with the addition of two active research scientists, Vivian Stannett and Michael Szwarc, the need for space became pressing. Baker Laboratory was in the planning stage and nearly the only available space was a sophomore organic laboratory. That course was transferred to Syracuse University and our graduate students carried out the old laboratory benches, painted the walls, and racks were set up for physico-chemical experimentation. At that time there was a great need for data for the packaging industry, and Vivian Stannett and Michael Szwarc obtained grants from the Technical Association of the Pulp and Paper Industry and the U.S. Quartermaster Corps. to study the permeability of plastic films and coated papers to gases. Soon the racks began to blossom vacuum systems and permeability cells and an abandoned dusty balance room was cleaned and received an infusion of quartz spring balances and cathetometers. The data that was gathered at that time is still of interest to people working in membrane science [3], a field that was virtually unknown and certainly of no practical importance at that time.

One day when I was looking up papers on wood hydrogenation, I came on a paper by Fierz-David that reported the formation of levoglucosan. I thought this looked like an interesting molecule and realized that I knew very little about its chemistry. On a library search, I learned that levoglucosan had been prepared in 1918 by Pictet, a Swiss chemist, by cellulose pyrolysis [4] and had been polymerized by him with acid catalysis to a multi-branched polymer [5]. This led us into a thirty year digression into the synthesis of stereoregular polysaccharides. It did not take very profound thought to realize if one blocked the free hydroxyl groups, it should be possible to form a linear polymer with acid catalysis. We learned that that had been attempted twenty years before by Irvine and Oldham [6]. They failed and initially so did we. However, a laboratory in Germany [7] and one in Latvia [8] found conditions that worked and we were able to demonstrate that the conditions used by the Latvian group led to stereoregular dextran [9]. If I could summarize about 30 years work, I might say we developed fairly general methods for making 1,2; 1,3; 1,4; and 1,6 anhydro sugar derivatives. The syntheses usually involved about nine steps to the perbenzylated monomer, the last of which was a ring-closure with base between a free hydroxyl group and a leaving group. Polymerization with a Lewis acid led to polymer. We prepared a number of 1,6-anhydro sugars, found optimum conditions for their stereoregular polymerization and copolymerization and determined the factors that determined their reactivity [10]. We also had similar good success with the 1,3-anhydro sugar derivatives that were previously unknown [11,12], and fair success with 1,2-anhydro sugars [13,14]. The 1,4-anhydro sugar derivatives were more difficult because in our hands, they polymerized to give α and β linkages and also mixtures of pyranose and furanose rings in the chains [15]. Professor Uryu at Tokyo University, however, has succeeded in preparing stereoregular pyranose and furanose polymers in the pentose series [16,17]. Our polymers have served as model compounds in immunology, allergy and enzymology and their use has clarified a number of structural and mechanistic problems. Professor Uryu's laboratory in Tokyo and Dr. Kobayashi in Professor Sumitomo's laboratory at Nagoya have been very active in this area and in the application of synthetic polysaccharides to biomedical applications. Professor Uryu will speak on this subject later in the Conference.

As chemistry department chairman in a forestry college, it became my responsibility to see that we carried out research in the application of chemistry to wood products, the latter a long established concentration at the College. Unfortunately, no one in our group knew much about adhesives or coatings and finishes. To attempt to perform research in those mature technologies was, for us, a sure trip into oblivion. My colleague, Dr. John Meyer, however, developed a method of treating wood with liquid monomers and a catalyst, and curing the product to form a wood-plastic combination that had many of the best properties of both [18,19]. Dr. Meyer's research led to the formation of a number of small wood specialty companies by entrepreneurs around the world who sought his advice and suggestions.

For my part I looked into the question of internal chemical treatments of wood. At first, the field did not seem to be an inviting prospect to me. Most of the work had a heavy technological component, but the problems seemed to be difficult to tackle scientifically, especially by an organic polymer chemist. Furthermore, outside funding was a necessity for any extensive studies and for that, some novel feature was certainly required. However, it occurred to me that chemical treatments had generally been carried out by impregnating wood with liquids or solutions, not an easy task. I decided that perhaps a gas phase treatment would be easier and would provide enough novelty for outside support.

The next step was to find out if any gases would produce any marked change in wood properties. A perusal of the Chemical Rubber Handbook gave a list of volatile compounds, mostly oxides, halides, and hydrides. Ammonia was at the top of the alphabetical list and immediately caught my eye. Many years before, a collaboration between A. J. Barry and F. C. Peterson at our college and Aden King at Syracuse University had shown that ammonia could enter cellulose crystallites [20]. More recently, I had established what properties were necessary to soften and swell lignin. It was obvious to me that ammonia would soften both. I asked Mary Burdick, a coworker, to collect some liquid ammonia in a Dewar flask and slip some wood samples into it. In an hour or two, Mary came back with some small wood samples bent into delicate curved shapes that she had formed with finger tip pressure. The wood had softened to about the consistency of stiff leather and when it was bent, unlike wood that has been steamed, it did not spring back to its original form. When the ammonia evaporated, the wood hardened again and appeared to be unchanged except for its shape [21].

Now it so happened that that year my colleague, Michael Szwarc, had been chosen as chairman of the Gordon Conference on Polymers. In order to highlight the College's polymer program, he invited me to speak on "The Future of Cellulose Research in Industry". I was not sure of my credentials for that assignment, but I went armed for the lecture with about 150 wooden tongue depressors that we had twisted into helices, coils, overhand knots, and diverse unnamed shapes not previously seen in wood. In the talk, I made a number of generalities about cellulose research and then proceeded to discuss the ammonia plasticization of wood. At the appropriate time, I passed the samples out to the audience and I believe they caused a sensation. I also showed that steam-bent wood, on treatment with hot water, would straighten out to nearly its original shape; whereas an ammonia-bent sample develops a permanent set, has forgotten its original form, and is nearly unchanged on hot water treatment. Subsequent developments may have hinged on that observation.

A scientist from the Norwegian Paper Research Institute in Oslo indicated to me that he was interested in testing the effect of ammonia on paper. That institute then developed an ammonia process for making extensible paper. Although that was not commercialized, the nearby Norwegian Textile Institute investigated the influence of liquid ammonia on cotton cloth. At a Cellulose Research Conference around 1981, F. W. Herrick reported that around the same time A. D. Little began to carry out research on the ammonia treatment of textiles and as a result of our little demonstration and publications, interest in the ammonia treatment of cotton was regenerated and from it was derived the ammonia-treatment of denims, shirts and sheetings and the Sanfor-Set process [22]. I hope he is correct. I believe the fact that we emphasized the possibility of temporary plasticization and permanent set may have renewed interest in the field. Another unexpected minor consequence of the work resulted when Dr. A. Zachariades left my colleague Michael Szwarc's laboratory to work with Professor Roger Porter at the University of Massachusetts. Dr. Porter was working on the cold drawing of polymers for increased orientation. The results with nylon were not particularly gratifying. Zachariades suggested a temporary plasticization with ammonia and was able to show a demonstrable improvement in the cold drawing process.

The interest in obtaining greater orientation of molecules in fibers has of course a long history. I remember that one of the first articles in polymer science that I read was one by Herman Mark discussing the marked difference between the theoretical strength of fibers, based on bond strengths and that achieved in practice. He related it at that time to disorder in molecular packing. I believe one of Carothers' papers makes the same point. Now with the spinning of polymeric liquid crystalline systems, these speculations are confirmed. Unfortunately we do not yet have a suitable liquid crystalline system for cellulose.

One of the newer approaches at cellulose solution was made several years ago by two of my former colleagues. One of my early students, Dr. Herman Marder, went on to a distinguished career in industry but remembered from his college days that there were no true solvent systems for cellulose, that effectively cellulose had to be derivatized to dissolve in reasonable concentrations. He was clever enough to disbelieve this, and the problem disturbed him, so when he was in a suitable position to do something about it, he hired Dr. Morton Litt as consultant to work on the problem. Litt was a very wise choice, for he is in my opinion one of the most versatile polymer scientists currently active. Litt made himself knowledgeable about cellulose solubility, made some speculative calculations based on solubility parameter and demonstrated that hydrazine formed true solutions of cellulose, that it dissolved cellulose at elevated temperatures and remained insoluble or as supercooled solutions at low temperatures [23]. Now hydrazine is a remarkable solvent for cellulose for it has a quite low boiling point of $113^{O}C$, so one could actually imagine dry spinning cellulose. However, that must remain in the imagination, since, as I am sure we all know, hydrazine also is highly explosive and probably carcinogenic. Nevertheless I considered the discovery a remarkable intellectual achievement for the time. Litt tells me he believes that the cellulose hydrazine system at higher concentrations is liquid crystalline but for obvious reasons he has not gone further with that research.

Dr. John Cuculo of North Caroline State University has also been investigating an interesting solvent system, a high concentration of ammonium thiocyanate in liquid ammonia. The work was reported at our recent Tenth Cellulose Conference in Syracuse and by Dr. Cuculo in a Symposium honoring Vivian Stannett at Raleigh, N. C. It appears that liquid crystalline cellulosic systems are formed in this medium and fibers can be regenerated in methanol and isopropyl alcohol. Preliminary results appear to indicate higher orientation and greater strength. It will be interesting to follow this work as more information becomes available.

However, I believe that those of us in academia are especially impressed with the investment industry has made and commitment it has shown in developing the new cellulosic fibers. I believe it is widely known that American Enka and its parent company AKZO have had a substantial effort toward the regeneration of cellulose from N-methylmorpholine oxide solution and that Courtaulds has announced a new fiber "Tencel" reconstituted from amine oxide solution. A friend who has seen this new fiber described it to me as absorbent with a crispness like cotton, excellent wash properties and modulus, about 5g/denier. Certainly this is an enormous technical advance that goes far toward solving the difficult environmental and recovery problems of the viscose process. I have received some data on the new fiber, kindness of Courtaulds, but I expect we will hear more about this development later in the program from an expert.

My friend Robert Marchessault still takes a fiendish delight in reminding me that I am not a fiber expert, only a misplaced organic chemist. Many years ago when Bob was a member of our faculty, Professor Donald Lundgren of the Biology Department of Syracuse University was working on the microbial formation of the β-hydroxybutyrate fiber. He enlisted Marchessault's aid in characterizing the supermolecular structure and properties of the fiber. Bob was extremely enthusiastic about this novel material and asked my opinion. I was well aware of the easy elimination reaction of β-hydroxybutyrate derivatives and visualized a β-hydroxybutyrate textile falling apart on the ironing board to the consternation of a housewife, and I opined that it would never make the grade as a commercial fiber. Now of course it is sold as a biodegradable surgical suture and the discovery that some microorganisms can incorporate related β-hydroxyacids in the polymer also permits the formation of a range of copolymers of varying properties. Never say never!

In the early 1960's we undertook a study of the reaction of ozone on wood and pulp, primarily with Z. Osawa, now professor at Gunma University. We published our results in 1963 [24], the same year that the Tappi Monograph on Bleaching of Pulp, edited by Professor Howard Rapson [25] was published. The emphases of the two publications were totally different. The Monograph dealt very thoroughly with the chemical reactions and the technology involved in pulp bleaching, while our publications emphasized the identification of the rate-determining step in the transport of ozone to reaction site, which was usually the movement of the reagent through an immobilized layer of water surrounding the fiber. To interpret the problem, we set up a simplified model of the reaction to describe the transport. I presented the results at a Cellulose Conference at Syracuse and a Wood Chemistry Symposium in the Laurentians. After one of the presentations, one of my friends said, "That was a good paper, Connie, but why don't you work with something practical, like chlorine". Howard Rapson also made a curious remark to the effect that if transport across an immobilized liquid layer was the problem, as a technologist he would have to solve it. I didn't understand his remark because I thought I had demonstrated the superiority of gas phase bleaching on moist pulp.

From reading his next publication [26] and his presentation at the next bleaching conference, I gather that we stimulated his thinking. It appears that probably he went back to Toronto, filled a glass tube with pulp and ran a series of bleaching solutions and washes through the pulp, and demonstrated a much faster bleaching rate by his method of "dynamic bleaching". Kamyr developed equipment for their displacement bleaching process based on the same principle, and an article in Tappi magazine reported that fifty percent of the new bleach plants were of the displacement type [27]. At the most recent International Bleaching Conference in Orlando, Florida, S. C. Pugliese and T. J. McDonough presented a sophisticated study of the rate-determining step in transport processes involved in the chlorination of pulp with high shear mixing [28]. They use a flow reactor for the analysis. Certainly the intellectual climate has greatly changed since 1963, and emphasis on transport processes is common today.

It is usually foolish to claim priority in fields of technical development, and I would not do so here. There are very few completely new ideas, and almost always someone earlier has had similar thoughts or approaches. That is certainly true in most of the fields I have studied. However ideas do have consequences. When expressed in a general form they can enrich the intellectual climate and can stimulate others to apply them in unexpected ways. That is one aspect that makes research a delight and a continuing adventure. Good Hunting!

REFERENCES

1. Schuerch, C., The Solvent Properties of Liquids and Their Relation to Lignin Solubility. J. Am. Chem. Soc., 1952, **74**, 5061-67.

2. Herrick, F.W. and Hergert, H.L., Utilization of Chemicals from Wood. In The Structure, Biosynthesis, and Degradation of Wood, eds. F.A. Loewus and V.C. Runeckles, Plenum Press, New York, 1977, pp. 443-515.

3. Stannett, V.T., Szwarc M., Bhargava, R.L., Meyer, J.A., Myers, A.W. and Rogers, C.E., Permeability of Plastic Films and Coated Papers to Gases and Vapors, TAPPI Monograph No. 23, TAPPI Press, Atlanta, GA. 1962.

4. Pictet, A. and Sarasin V., Distillation of Cellulose and Starch in Vacuo. Helv. Chim. Acta, 1918, **1**, 78-96.

5. Pictet A. Transformation of Levoglucosane into Dextrin. Helv. Chim. Acta, 1918, **1**, 226-30.

6. Irvine, J.C. and Oldham, J.W.H., Polymerization of ß-Glucosan. J. Chem. Soc., 1925, **127**, 2903-22.

7. Hutten, U., Synthesen von Oligo- und Polysacchariden Dissertation, Technische Hochschule, Stuttgart 1961.

8. Korshak, V.V., Golova, O.P., Sergeev, V.A., Merlis, N.M., and Pernikis, R.Ya., Vysokomol. Soedin, 1961, **3**, 477-85.

9. Tu, C.C. and Schuerch, C., The Stereospecificity of Trimethyl Levoglucosan Polymerization. Polymer Lett. 1963, **1**, 163-65.

10. Schuerch, C., Synthesis and Polymerization of Anhydro Sugars. In Advances in Carbohydrate Chemistry and Biochemistry, eds. R.S. Tipson and D. Horton, Academic Press, New York, 1981, vol. 39, pp. 157-212.

11. Kong, F. and Schuerch, C., Synthesis of (1→3)-α-D-Mannopyranan., Macromolecules, 1984, **17**, 983-89.

12. Good, F.J. Jr. and Schuerch C., Synthesis of (1→3)-α-D-Glucopyranan., Macromolecules, 1985, **18**, 595-99.

13. Uryu, T., Harima, K., Tsuruta, T., Suzuki, C., Yoshino, N., and Schuerch, C., Ring-Opening Polymerization of 1,2-Anhydro-3,4,6-tri-o-benzyl-ß-D-Mannopyranose, J. Polym. Sci.:Polym. Chem. Ed., 1984, **22**, 3593-98.

14. Trumbo, D.L. and Schuerch, C., Steric Control in the Polymerization of 1,2-Anhydro-3,4,6-tri-O-benzyl-ß-D-Mannopyranose, Carbohydr. Res.,

1985, **135**, 195-202.

15. Kops, J. and Schuerch, C., The Polymerization of 1,4-Anhydro-Sugar Derivatives, J. Polym. Sci., Part C Polym. Symp., 1965, **11**, 119-138.

16. Uryu, T., Yamanouchi, J., Kato, T., Higuchi, S. and Matsuzaki, K., Selective Ring-Opening Polymerization of 1,4-Anhydro-α-D-ribopyranoses, J. Am. Chem. Soc., 1983, **105**, 6865-71.

17. Uryu, T., Yamanouchi, J., Hayashi, S., Tamaki, H., and Matsuzaki, K., Synthesis of Stereoregular (1→5)-α-D-Xylofuranan, Macromolecules, 1983, **16**, 320-26.

18. Meyer, J.A., Wood-Polymer Materials, State of the Art, Wood Science, 1981, **14**, No. 2, 49-54.

19. Meyer, J.A., Industrial Use of Wood-Polymer Materials, Forest Products J., 1982, **32**, No. 1, 24-29.

20. Barry, A.J., Peterson, F.C., and King, A.J., Interaction of Cellulose and Liquid Ammonia, J. Am. Chem. Soc., 1936, **58**, 333-37.

21. Schuerch, C., Plasticizing Wood with Liquid Ammonia, Ind. Eng. Chem., 1963, **55**, 39.

22. Herrick, F.W., Review of the Treatment of Cellulose with Liquid Ammonia, J. Appl. Polym. Sci., Appl. Polym. Symp., 1983, **37**, 993-1023.

23. Litt, M.H. and Kumar, N.G., Cellulose Solutions and Products Prepared Therefrom, U.S. Pat. 4,028,132, June 7, 1977.

24. Osawa, Z. and Schuerch C., The Action of Ozone on Cellulosic Materials, TAPPI, 1963, **46**, No. 2, 79-89.

25. The Bleaching of Pulp, ed. W. Howard Rapson, Tappi Monograph, No. 27, TAPPI Press, Atlanta, Georgia, 1963.

26. Rapson, W.H. and Anderson, C.B., Dynamic Bleaching, TAPPI, 1966, **49**, No. 8, 329-34.

27. Reeve, D.W., The Future of Bleaching, TAPPI, 1985, **68**, No. 6, 34-37.

28. Puglies, S.C. and McDonough, T.J., Kraft Pulp Chlorination, A New Mechanistic Description, I.P.C. Technical Paper Series, No. 282, Institute of Paper Chemistry, Appleton, Wisconsin, 1988.

DISCUSSION

chaired by J. Nakano (Fukui Inst. Tech)

Q: J. Nakano (Fukui Inst. Tech.)
In your lecture there were not too many comments about lignin. In the future, is it possible to use lignin as polymer or as fine chemicals ?

A: The structure of lignin is much too complex to serve as a source of fine chemicals economically. Too large a product mixture results from lignin degradation, many of the products in all probably will not be of significant market value. Almost all successful applications have been for lignin and lignin derived materials. Except for minor exception - like vanillin and DMSO I expect this situation to continue.

Q: A. Blažej (Slovak Tech. Univ.)
(1) How do you prepare anhydroglucose - it is practically levoglucosan - was it prepared by pyrolysis (by thermal pre-treatment) ?
(2) It is used like a monomer, how do you make the pre-treatment for polymerization ?

A: Levoglucosan was prepared both by pyrolysis and by multi-step synthesis from glucose. Pyrolysis requires high purity cellulose and special equipment for high yield. Chemical synthesis is more readily controlled in the laboratory.
Levoglucosan must be etherified - usually benzylated for cationic polymerization. The benzyl ether groups are removed to give the free polysaccharide.

Q: J. J. Silvy (Ec. Franc. Papeterie)
Hydrophilicity of cellulosic fibers is a natural property of this material and we need of it for products with high absorbency and to beat the pulp for paper manufacturing in good conditions without to expend too much energy.
On the contrary, hygrosensitivity of cellulosic products is sometime undesirable because it causes unstability of the paper properties - loss of strength and curl of paper. Have you some idea of the magic way we must go to control these two opposite compartments in order to increase the use of cellulosic materials in the future ?

A: Hydrophilicity, of course, can be increased by any factor that increases disorder in the supermolecular structure, swelling agents, lower DP fractions etc. This has been studied on effects on beating etc. However too much hydrophilicity results in deterioration of many useful properties, as you state. The forces involved between water and cellulose are of course very strong. Surface treatments or coating or use of composites at best slow the sorption of water but do not completely inhibit it. Any solution to these problems would appear to have to be a practical compromise, not a complete solution.

FUTURE PROSPECT OF ABSORBENCY

Pronoy K. Chatterjee
Absorbent Technology, Johnson & Johnson
21 Lake Drive, East Windsor, N.J.   08520

SYNOPSIS

Future trend of the technology has been discussed with a
critical analysis of the current state of the art on theories
of absorbency, test methods and materials. Prospect of new and
functionally improved products would depend upon the
development of modified theories, predictive models and new
composites with few interfaces between components constructed
by pad stabilization technique and new superabsorbent systems.
Environmental and safety concerns on chlorine bleached pulp and
new materials developed through biotechnology are likely to
impact absorbency technology.

INTRODUCTION

Projection of the future of the absorbency field based on an
analysis of the technology gap that exists in the area of fluid
absorbency in fibrous media is the theme of this presentation.
The discussion will be mostly confined to the basic scientific
aspect of absorbency and absorbent materials, not on specific
design features of any commercial product. More specifically,
the following items will be discussed: theories of absorbency
and predictive models, fluid characteristics, limitations of
current test methods, progress on composites, superabsorbent
development trend, new cellulosic fibers and the role of
biotechnology.

THEORIES OF ABSORBENCY AND PREDICTIVE MODELS

Many classical theories on fluid flow in porous media [1],
which were being originally developed for soil science, have
been adapted to explain the liquid absorption phenomena in
fibrous absorbent materials. Examples are: Darcy's law,
Kozeny-Carmen equation, network model, drag theory, semi-
infinite radial flow and so on. These theories are not
adequate to address the situation where a porous structure is
spontaneously modified as it interacts with the fluid and the
flow characteristic undergoes a continual change as the fluid

begins to get transformed due to environmental factors.
Application of superabsorbent which immobilizes the flow of
fluid and frequently modifies the pore structure further adds
to the complexity of the phenomenon. There is a definite need
for modified models to understand the flow of fluid in fibrous
media.

Lucas-Washburn equation [1], which has been widely used in the
absorbency field to interpret the liquid wicking phenomena, was
derived for liquid rise in an ideal capillary system. There
has been controversies regarding its applicability in low
density fiber composites where the so called "capillaries" are
inter-connected and they tend to collapse or expand on
absorbing liquid. Many suggestions for the modification of
Lucas-Washburn equation have been recently proposed which may
lead to the development of a more appropriate wicking theory
for absorbent products [1,2,3].

Determination of liquid flow pattern inside the absorbent
composite is vitally important to develop a predictive model
and to understand the flow characteristics through different
elements as well as interfacial regions. The flow of liquid at
the interfaces and boundaries is least understood. All
absorbent products have discrete elements such as nonwoven
facing, absorbent core, high density - low density regions,
tissue, barrier film, etc. The presence of these discrete
elements creates numerous interfacial regions with
characteristic properties. When fluid impinges on the facing
material, it transfers to the next absorbent layer through the
interface of the facing and the intermediate transfer layer.
If the facing is bonded to the transfer or absorbent layer, the
migration takes place quite rapidly. If it is detached, air
barrier would develop at the interface and the fluid may roll
over to the side. This type of uncontrolled flow occurs not
only at the interface of the transfer and absorbent layers but
at numerous other locations within the entire absorbent
structure. Therefore, experimental determination of fluid flow
pattern under variety of conditions is essential to develop a
meaningful predictive model which would lead to the development
of more effective absorbent articles.

Recently, it has been reported that certain medical diagnostic
tools, such as X-ray computed tomographic scanner (CT scanner)
or magnetic resonance imaging (MRI), can provide qualitative as
well as quantitative information of the liquid distribution in
an absorbent structure [3]. As more data are generated, it is
possible that a statistical fluid flow model can be developed
through these techniques. Ultimately with the help of an
appropriate computer software and utilizing the model for
prototypes and their performances one can simulate new product
designs on a computer screen. Thus, the future product design
will heavily depend on computer aided design skills.

FLUID CHARACTERISTICS

Most absorbent products in the market are for absorbing body fluids, e.g., urine, blood or menstrual fluid. Unlike water, the characteristic properties of all body fluids change with time and environmental conditions. The continual change of the chemical nature of the fluid results in a constant change of the flow mechanism.

There are several reviews in the literature [1] which describe the rheology of blood. The data on menstrual fluid which is even more complex are not available in the literature. Blood and menstrual fluid both are thixotropic in nature, but blood frequently coagulates in contact with a foreign material whereas menstrual fluid does not. The structure of the proteinacious material present in menstrual fluid would undergo drastic changes in presence of a surface active agent. Such changes would certainly alter the flow properties of blood and menstrual fluid. Understanding of these phenomena would have profound effect on product design, test methods and predicative mathematical or statistical models.

LIMITATIONS OF CURRENT TEST METHODS

There are numerous published in-vitro test methods for absorbent products which are crude, subjective and perhaps designed to meet the need of ranking different products, not for scientifically characterizing the products for actual performance. The most frequently used test procedure for evaluating products is based on real life use or in-vivo technique. The in-vivo test procedure which costs thousands of dollars is primarily dependent on responses to questionnaires prepared by individuals. There is no industry wide standard system or protocol for these tests and they are very subjective.

For diaper type of products, the Swedish Textile Research Institute has developed a test procedure, which is known as TEFO tester [4] and it has been used by many industries to characterize their products. For napkins and other absorbent products no such method is available or acceptable industry wide. As the need has been recognized, it is expected that the science and technology of testing products would significantly progress in the near future.

Regarding methods to characterize individual components of absorbent products, we see a proliferation of techniques which range from simple to quite sophisticated ones [1,3,4]. They do characterize certain attributes of composites in terms of their interactions with simple liquids but not always correlate with the performance of finished products. Such a correlation is going to evolve as the scientific knowledge on composites and fluids is further developed.

ABSORBENT COMPOSITES

Wood pulp as an absorbent core suffers from its limitations in terms of fluid transporting ability and collapsibility in contact with the fluid, unless it is compressed to a high density level. Blending with various kinds of textile fibers may overcome some of these limitations, but still major innovation is required to meet the desired criteria. With the advent of thermoplastic fusible fibers which on heating stabilize the pulp structure, major innovations are expected in custom manufacturing of unitized, more stable absorbent composites. Also composites with thermoplastic hydrophilic foam, if and when it is commercially developed with affordable price, will play a significant role in absorbent products category. An ideal unitary structure would eliminate separate facing material and thus avoid uncontrolled fluid flow at the interface. This type of uncontrolled flow arises when facing physically separates from the absorbent core during the use.

Most absorbent composites have some kind of nonwoven facing materials which separate the absorbent core from the environment. It has been reported in the literature that perforated plastic film instead of conventional fibrous nonwoven provided a drier feel. However, plastic film is aesthetically less desirable than the nonwoven fabric. This paradoxical situation could lead to the emergence of new nonwoven fabric with modified fiber surface to enhance the dryness without sacrificing the aesthetics.

Prefabricated composites with localized distribution of bonded superabsorbent polymer is a potential candidate for the future. With the advent of new types of synthetic fibers [1], hollow and resilient, resilient as well as hydrophilic, bicomponent, thermoplastic, bioactive and others, the absorbent composites are likely to get more sophisticated as well as more effective.

SUPERABSORBENT DEVELOPMENT TREND

Except a few industries, most are now producing granular polymer from acrylic acid. Japan, being in the forefront of superabsorbent market, is already experiencing shortage of acrylic acid. Some acrylic acid producers in Japan have been increasing their production capacity [5]. If the current trend of increasing superabosrbent consumption continues, soon the superabsorbent manufacturers will have to use other monomers or a different type of superabsorbent system. This situation may result in the revival of the cellulose based superabsorbent which could not make its place in the market primarily because of economic reasons. Unlike granular synthetic superabsorbent, a fiber based superabsorbent, e.g. grafted cellulose copolymer [1] would provide improved capillarity on interacting with liquid due to the unfolding of cellulose lamella and due to higher overall structural rigidity. This improvement in capillary structure is expected to accelerate the swelling or liquid imbibition speed rather than decreasing the speed as is the case for synthetic acrylic based superabsorbent polymer.

NEW CELLULOSIC FIBERS

Bleached kraft softwood pulp is by far the most widely used cellulosic fiber in absorbent products. The recent concern on dioxin [6] in kraft pulp and chlorine compounds in the effluent of pulp mills has led Sweden to switch from bleached chemical pulp in most of their absorbent products to bleached chemco-thermomechanical pulp (CTMP) where bleaching was done with hydrogen peroxide instead of chlorine and chlorine dioxide. Even though the amount of dioxin present in kraft pulp is way below the established toxic range, this move of Sweden may lead other countries to use CTMP. However, the high energy requirement for the production of CTMP will have a significant negative effect on meeting its future worldwide demand. It is also expected that a more economical process for bleached kraft pulp will be developed where bleaching would be done with substantially reduced amount of chlorine compounds or with alternate bleaching agents. Whatever way the pulp market moves, the properties and/or cost of pulp will certainly change. This change will inevitably lead to a significant modification of the absorbent products design to make it more efficient in terms of its total utilization of pulp during the use. Alternate raw materials such as unbleached pulp and non-wood fibers may also find their place in absorbent products category.

BIOTECHNOLOGY IN ABSORBENCY

Like in most other fields, biotechnology will have an impact on absorbent products field too. Biodegradable polymers such as polyhydroxy buterate and its copolymer produced by fermentation technique or starch based biodegradable copolymer could find their place in absorbent products in the future. If absorbent product has to be totally biodegradable by legislation, all its components such as barrier film or facings have to be also biodegradable.

Also several biochemical processes are being developed to produce cellulose e.g., use of certain bacteria such as acetobacter xylinum or cloning enzymes which produce cellulose in trees and plants [3]. The removal of lignin by a bacterial process is being studied in certain institutes and industries. These processes could alter the property of pulp and thus its interactions with fluid. Through biotechnology, other new materials or old materials with different properties will be developed and those will take a significant role in future absorbent products.

CONCLUDING REMARKS

The prospect of future absorbent products is dependent on more thorough understanding of the material - fluid interactions and the development of innovative concepts on composites. The progress on high speed, more versatile and flexible machineries will also contribute to the advancement of the technology. More active research collaboration of industries and academia is needed to meet the objective at a faster pace.

REFERENCES

1. "Absorbency", Edited by P.K. Chatterjee, Elesevier Science Publisher, Amsterdam, The Netherlands (1985).

2. Salminen, P.J., TAPPI, 71, p. 195, (1988).

3. Proceedings of the Symposium on Nonwovens, Absorbency/Testing Section, Third Chemical Congress of North America, Toronto, Canada, June 5-11, 1988, American Chemical Society, U.S.A., In Press; Abstract in Winter Newsletter, Cellulose Paper and Textile Div., American Chemical Society, September 1987, p. 112-116, 139-143.

4. Proceedings of 1987 International Dissolving Pulp Conference, Geneva, Switzerland, March 24-27, TAPPI Press, Atlanta, Ga., U.S.A., (1987), p. 211-268.

5. Ohmura k., Nonwovens Industry, U.S.A., p. 30, August 1980.

6. News Article in Pulp & Paper Week, U.S.A., p.9, April 25, 1988.

DISCUSSION

chaired by V. T. Stannett (North Carolina State Univ.)

Q: M. Sakamoto (Tokyo Inst. Tech.)
I would like to ask you on superabsorbent polymers. Superabsorbent polymers are often regarded as newly develop-ed, high-tech materials. However, the current end-uses for them are high-volume, throwaway products and so the cost is the key factor for the commercial success. It seems to me

that to find high-tech end-uses for superabsorbent polymers is important for stimulation of R & D efforts for super-absorbent polymers because high-tech end-use products can absorb the initial high cost. Do you have any ideas about the high-tech end-uses ?

Another related question is on the prices of diapers, napkins etc. Do you think some of the consumers would like to pay more for the better performance for such products with eq. cellulosic superabsorbent fibers ?

A: To answer your first part of the question, yes. I agree that high volume-throwaway products are very cost sensitive and therefore the application of superabsorbency is always a problem. However new innovative design of those products might allow us to use the high cost material and still the products could be economical. I do not have any specific idea of current high tech products where high cost superabsorbency could be used. However, there are many high cost surgical dressings which are expensive where one might consider an application of superabsorbency.

To answer the second part of the question, if the fine benefit is there and if the consumer perceives the benefit, I believe they will pay the higher price.

Q: R. Narayan (Purdue Univ.)
The effect of zero gravity on absorption of fluids spe-cifically in the space station and how it relates to models in the area ?

A: Certainly the model would need revision when the product is used in space. However, at this point it is difficult to comment on the exact effect.

Q: V. T. Stannett (North Carolina State Univ.)
How do the maximum sorbencies of cellulose based super-absorbency(Cell SA) compare with the new synthetics(Syn SA) ?

A: If tested with 1% salt water solution, the ratio of Cell SA versus Syn SA would be approximate 1.5 to 2.0.

NEW MATERIALS FROM CELLULOSE OR LIGNOCELLULOSE SOLUTIONS IN AMINE OXIDE SYSTEMS

ALAIN PEGUY

Centre de Recherches sur les Macromolécules Végétales,
CNRS, B.P. 53 X, 38041 Grenoble cedex, France

ABSTRACT

We have studied in our laboratory the dissolution of cellulose or lignocellulose in N-methylmorpholine N-oxide and in dimethylethanolamine N-oxide. To dissolve polysaccharides the temperature and water content must be controlled. The phase diagram of the ternary system MMNO/water/ cellulose was established. Cellulose or lignocellulose solutions can be shaped into different forms. We have studied the spinning into fibres and the casting into films of the solutions. We have also experimented with several ways to get cellular structure from cellulose solutions in MMNO. We will compare the products obtained (fibers, films and sponges) with those obtained from other processes. Finally we will discuss the possibility and the advantage of mixing cellulose with other poly-saccharides or synthetic polymers.

INTRODUCTION

Cellulose is globally very important because it is highly abundant in nature. With cellulose we can make hydrophilic fibers for the textile industry but the main use for cellulose is in the papermaking industry.

To make textile fibers (rayon) the lack of a true solvent made it necessary in the past to derivatize cellulose to allow the obtention of a cellulose solution and make fibres, films or sponges.

Up till now these materials have been made with the viscose process which is slowly being abandoned in the western world due to its complexity and the pollution it causes. Moreover the viscose process does not allow the use of all kinds of cellulose as a raw material. Dissolving pulp must be used.

To avoid all these problems, research devoted to replacing the viscose process for making cellulose materials has occupied several laboratories around the world for several years. For the most part, the research was oriented towards a solvent system for cellulose.

Some years ago in our institute, our interest turned towards the use of tertiary amine oxides to dissolve polysaccharides. Some patents have demonstrated the power of tertiary amine oxides to dissolve great amounts of cellulose [1-3].

RESULTS AND DISCUSSION

These solvents have been demonstrated to be advantageous for fibers, with a high recovery rate for the solvent [4]. In our laboratory we have used extensively N-methylmorpholine N-oxide (MMNO) and dimethylethanolamine oxide (DMEAO). These two products (in their dissolving form) are crystalline at room temperature ; this is why any dissolution occurs at high temperatures (120°C).

A DSC trace of a MMNO sample shows that three forms exist :
- the anhydrous form, which is a very good solvent but has a melting temperature of 174°C (temperature to avoid on account of explosion risk) ;
- the monohydrate which melts at 74°C and is normally used to dissolve polysaccharides ;
- the so-called dihydrate (2.5 water molecules for 1 MMNO) is not a solvent of cellulose.

A crystallographic study has shown that the N → O group can form two hydrogen bonds. Combining crystallographic investigation and computer simulation leads to an understanding of the cellulose dissolution mechanism and provides information at a molecular level about the complexation of cellulose chains in MMNO. MMNO can be characterized as a structuring, stiffening and protective agent.

The study of the structure of DMEAO has shown three independant crystallophilically molecules. One of them is a pseudo-cyclic conformation stabilized by a strong intramolecular hydrogen bond. Using IR and NMR we may conclude that this conformation is the preferential one in solutions.

These different results allow us to explain why this linear tertiary amine oxide is a good solvent for cellulose.

Both solvents are used with a certain amount of water (roughly the monohydrate content). This water is essentially useful for decreasing the melting point and consequently the temperature of dissolution. But too

much water can prevent the dissolution so in order to have a good understanding of the dissolution and coagulation conditions of cellulose, we established the phase diagram of the ternary system MMNO/water/ cellulose. The dissolution process and the coagulation in the cellulose can be controlled just by adjusting the water content and the temperature.

We have studied the crystallization of cellulose solutions and we have shown that cellulose does not crystallize and is textured by the growth of solvent crystals. From this we can prepare microporous cellulose membranes for example. The structure of these materials will depend on parameters such as the melting temperature, the crystallization temperature, the viscosity (also related to the degree of polymerization) and the rate of crystallization.

A rheological study was undertaken and solutions were found to be strongly viscoelastic [5].

High concentrations of cellulose give anisotropic solutions. Optical and transmission electron microscopy have confirmed the orientation of cellulose chains in fibers after shearing of an anisotropic cellulose solution [6].

The cellulose solutions (roughly 15% (w/w)) are spinnable. We have used the dry jet wet spinning technique. The cellulose was coagulated in water.

The fibers obtained from MMNO solutions have mechanical properties similar to what is obtained with the best fibers spun from viscose solutions.

A study of the influence of spinning process parameters is being done in collaboration with COURTAULDS. We have determined the fiber's structure using X-ray, optical and electron microscopy.

We also built a piece of equipment to measure the birefringence of the fibers on line. With this apparatus we can follow the alignment of cellulose chains along the pathway in the air gap. We can also have an access to the rate of regeneration of cellulose in the coagulate fiber.

The use of cellulose solutions in MMNO gives access to new fibers. Indeed their structure and some properties are different. But these solutions may lead to other materials.

Recently our interest was turned towards the development of a new process to obtain cellular cellulose (like sponges). Until now artificial cellulose sponges are made through the viscose process. Sponge manufacturers would like to have a much more versatile process in the future.

We have experimented with several ways to get a cellular structure from cellulose solutions in MMNO. One key factor in the new process is the water content. The process, even straightforward, is governed by a number of parameters that we have studied. The materials obtained from MMNO solutions are similar to those from viscose solutions.

Viscose solutions also allow the manufacture of films (cellophane films) ; it is the same with MMNO-cellulose solutions. One important advantage in using the MMNO process, however, is the possibility to use all kinds of cellulose as a raw material, which is not the case for the viscose process where dissolving pulp must be used.
Taking advantage of this possibility, we have used lignocellulosics as starting materials to make films with the MMNO process. Since these products are cheap, this could be a way to decrease the cost of such films.

To obtain these lignocellulosics, advantage may be taken of the steam-explosion process [7] which can convert lignocellulosic material (like wood chips) into a reactive and finely divided product soluble in organic solvent such as tertiary amine oxides [8]. The conditions for the preparation and spinning of exploded wood solutions have been described. The study of the structure of these fibers was taken up and the mechanical properties were presented [9].

The use of the exploded wood/tertiary amine oxide system leads to the obtention of new biodegradable films. These films may have a natural colour from brown to black. The mechanical properties were studied and a comparison with a black commercial synthetic film has been made. The structure of the film was studied as well as the role of lignin as an antioxidant.

The peculiarity of MMNO to be able to dissolve synthetic polymers or polysaccharides can be used to make mixtures with cellulose and give rise to new products or new processes. Some preliminary results in this field will be given.

REFERENCES

1. Johnson, D.L., Br. Pat. n° 1,144,084, 1969.

2. Franks, N.E. and Varga, J.K., U.S. Pat. n° 4,145,532 (1979).

3. Franks, N.E. and Varga, J.K., U.S. Pat. n° 4,247,688 (1981).

4. Layman, P.L., Chemical and Engineering, 1987, 65, 22, p. 9-13.

5. Navard, P., Haudin, J.M., Quenin, I. and Péguy, A., Shear rheology of diluted solutions of high molecular weight cellulose. J. Applied Polym. Sci., 1986, 32, 5829-39.

6. Chanzy, H., Péguy, A., Chaunis, S. and Monzie, P., Oriented cellulose films and fibers from a mesophase system. J. Polym. Sci.Polym. Phys. Ed., 1980, 18, 1137-44.

7. DeLong, E.A., Canadian Pat. n° 1,141,371 (1983).

8. Chanzy, H., Paillet, M. and Péguy, A. Spinning of exploded wood from amine oxide systems. Polym. Comm., 1986, 27, 171-2.

9. Chanzy, H., Paillet, M., Péguy, A. and Vuong, R., Dissolution and spinning of exploded wood in amine oxide systems. In Wood and Cellulosics, Ellis Horwood Limited, Chichester, 1987, pp. 573-9.

DISCUSSION

chaired by R. H. Atalla (Inst. Paper Chem.)

Q: S. U. Hossain (Kimberly-Clark Co.)
The experiments you did with the exploded material showed a DP range of 1500 ~ 210. What was your raw material ?   Did you explode wood chips or wood pulp ?

A: We used wood chips as raw material.

Q: W. Berger (Tech. Univ. Dresden)
If you will explain the dissolution process from cellulose with new solvent system with MMNO, you need exact information about the intermolecular interaction and solvation state. What is the influence of 1 mol H_2O in your system ? What is important for mixing with synthetic polymer in the same solvent systems ?

A: First of all, I would say that the best solvent for cellulose is the anhydrous form of MMNO. But the melting temperature of this crystalline form is too high to be used in order to avoid rapid decomposition and explosion. So we have to use the monohydrate (or a water content around 13.3%). We

have shown through the crystallographic studies that there is
an extensive hydrogen bonding scheme in the monohydrate crystal and so the use of MMNO monohydrate will not have for only
effect to decrease the dissolution temperature. For blends
with synthetic polymers the use of monohydrate MMNO will lead
to some interactions between MMNO, both polymers and water
(through hydrogen bonds).

RECENT TRENDS IN FINISHING OF CELLULOSIC FIBERS

HARRO PETERSEN

BASF Aktiengesellschaft, D 6700 Ludwigshafen am Rhein

Federal Republic of Germany

ABSTRACT

The relationships between crosslinking agents and properties of finished cellulose-containing fabrics have been studied with kinetically, thermodynamically and molecular modelling methods in combination with modern methods of D-optimal designs. With extremely modified hydroxymethylated 4.5-dihydroxy-ethylene ureas it is possible to reach optimum crosslinking effects with a common optimum of all desired levels of properties, including a minimum of formaldehyde liberation during the finishing process and on finished fabrics.

INTRODUCTION

The shrinkage and creasing of cellulosic fibers depend on the dimensional changes in the fiber during the transition from the dry to the wet stage. These disadvantages can be reduced by crosslinking the hydroxyl groups of cellulose molecules. At the same time the tensile strength of the fiber will be diminished together with a loss in abrasion resistance. The main difficulty in finishing cellulose-containing fabrics is to keep these losses within tolerable limits to impart good easy-care properties.

The most important requirements for finishing cotton fabrics and other cellulose-containing textiles are a controllable reactivity, a high level of durable press ratings, high dry and wet crease recovery angles, a low shrinkage and a low loss of abrasion, tensile and tear strength, a good hydrolysis stability of the finishing agent and the finished fabric, and no effect on shade and fastness properties of dyed fabrics. Another very important property is a low release of formaldehyde during the application and curing procedure and afterwards on sensitized or finished fabrics. In practice only the N-hydroxymethyl and N-alkoxymethyl compounds of acyclic and cyclic ureas, carbamates, carboxylic acid amides and aminotriazines are employed.

REACTANTS WITH LOW RELEASE OF FORMALDEHYDE

Fore several years the most important crosslinking agents have been based on hydroxymethylated and alkoxymethylated 4.5-dihydroxy- and 4.5-dialkoxy-ethylene ureas. On cellulose-containing textiles it is possible to achieve with this class of finishing agents high durable press ratings, low levels of shrinkage, high dry and wet crease recovery angles and a good hydrolysis stability. A new problem in the textile finishing industry is the release of formaldehyde during impregnation with the solution of formaldehyde-containing crosslinkers, drying and curing and the release of formaldehyde on the finished fabrics.

N-Hydroxymethyl and N-alkoxymethyl compounds are equilibrium mixtures and contain small amounts of free formaldehyde. During impregnation of the fabrics with the finishing pad bath, the free formaldehyde is transferred to the fabric, where some of it is discharged with the exhaust air in drying and curing. Additional formaldehyde may be liberated by partial hydrolysis in the curing process and during storage of finished textiles. The amounts of formaldehyde in the solutions and the release of formaldehyde during the finishing process and during storage of the finished textiles depend on the chemical constitution of the finishing agent, the concentration, molar ratio between the amino group-containing compound and formaldehyde, and the temperature. Furthermore the absolute content of free formaldehyde in the finishing agent or finishing pad bath, the hydrolysis stability of the finishing agent and the crosslinked cellulose, the type and amount of catalyst, the pH value on the fabric, and the application conditions are important factors for the release of formaldehyde (1). Three main facts are necessary to obtain finishing agents with an extremely low content and release of formaldehyde.

$$- \overset{\overset{O}{\parallel}}{\underset{|}{N}} \quad \overset{}{\underset{|}{N}} - CH_2OH \rightleftharpoons - \overset{\overset{O}{\parallel}}{\underset{|}{N}} \quad \overset{}{\underset{|}{N}}H + CH_2O$$

$$- \overset{\overset{O}{\parallel}}{\underset{|}{N}} \quad \overset{k_1}{\underset{|}{N}} \vdots CH_2 - OR + H_2O \rightleftharpoons - \overset{\overset{O}{\parallel}}{\underset{|}{N}} \quad \overset{}{\underset{|}{N}}H + HOCH_2OR$$

$$- \overset{\overset{O}{\parallel}}{\underset{|}{N}} \quad \overset{k_2}{\underset{|}{N}} - CH_2 \vdots OR + H_2O \rightleftharpoons - \overset{\overset{O}{\parallel}}{\underset{|}{N}} \quad \overset{}{\underset{|}{N}}H + HOCH_2OR$$

An extremely low level of free formaldehyde at the equilibrium state, a very low rate constant k_1 for the hydrolysis of the N-C-bonds and a low rate constant k_2 for the hydrolysis of the C-O-bonds are important for the synthesis of finishing agents with these desired properties. The constants k_2 for the cleavage of the C-O-bonds are not only a measure of the hydrolysis stability; these constants are at the same time a measure of the reactivity. The positions of the equilibria in aqueous solutions of N-hydroxymethyl compounds exhibit an extraordinarily pronounced dependence on concentration and temperature (2).

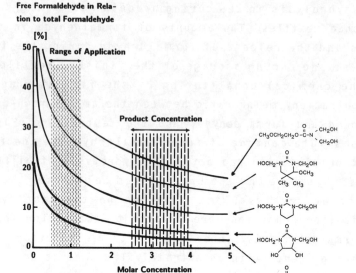

Figure 1. Equilibrium States of Dihydroxymethyl Compounds in Relation to Concentration at 20° C

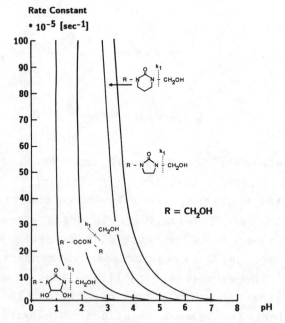

Figure 2. Hydrolysis Rate Constants for the first Hydroxymethyl Group at 70° C in Relation to pH

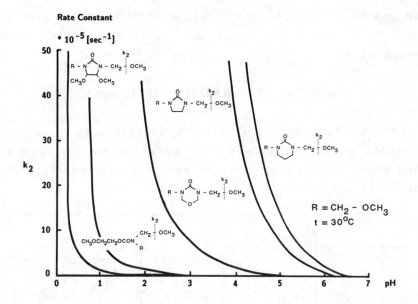

Figure 3. Rate Constants for the Cleavage of the C-O-Bond of the first Methoxymethyl Group at 30° C in Relation to pH

From Figure 1 it can be seen that the hydroxymethylated compounds of the 5-membered cyclic ureas show the best equilibrium states with the lowest levels of free formaldehyde (2). Figure 2 shows the hydrolysis rate constants k_1 for the cleavage of the first hydroxymethyl group of various dihydroxymethyl compounds in relation to the pH value at 70° C (2). From the point of view of the equilibrium state and the hydrolysis stability the hydroxymethylated 4.5-dihydroxy-ethylene urea behaves best. The third important property for a crosslinking agent with an extremely low content and release of formaldehyde is a very low rate constant k_2 for the hydrolysis of the C-O-bonds. As can be seen from Figure 3 the alkoxymethylated 4.5-dialkoxy-ethylene ureas show the best resistance to acid hydrolysis (2). Under alkaline conditions nearly all N-alkoxymethyl compounds are stable. Because of the identity of reactivity and hydrolysis stability of the C-O-bond, only slow-reacting crosslinking agents are suitable for finishes resistant to hydrolysis. The equilibrium positions and the hydrolysis stability of the N-C- and the C-O-bonds depend mainly on the electron densities at the nitrogen atoms.

GEOMETRY OF 5-MEMBERED CYCLIC UREAS

With the semi-empirical MINDO methodology it is possible to calculate the geometry of all these molecules by minimizing their total energies with respect to the corresponding geometrical variables.

Formula \ Torsion Angle	1 2 3 4	2 3 4 5	1 3 5 4	1 2 5 4	1 2 3 5
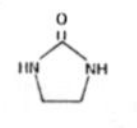	0.03	0.27	-0.43		-0.37
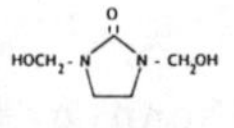	0.84	-0.95	0.65	-0.18	-0.38
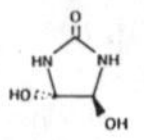	-6.2	10.72	-9.97	7.92	-1.86
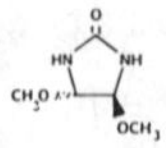	-0.19	2.72	-3.87	4.51	-2.95
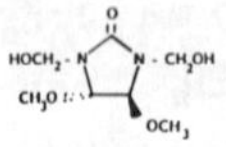	-6.88	15.33	-16.88	14.57	-5.67

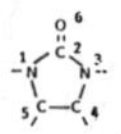

Formula \ Torsion Angle	1 2 3 4	2 3 4 5	1 3 5 4	1 2 5 4	1 2 3 5
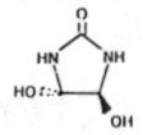	-6.2	10.72	-9.97	7.92	-1.86
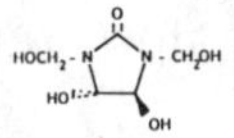	-6.78	13.14	-13.70	10.94	-3.25
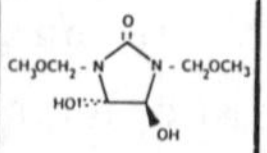	-5.77	12.79	-14.13	11.99	-4.54
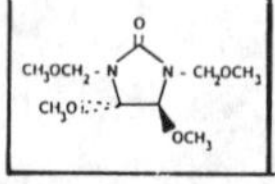	-6.77	15.22	-17.05	14.51	-5.65

Figure 4. Torsion Angles of Five-membered Cyclic Ureas

Figure 4 contains the torsional angles of the five-membered ring atoms for ethylene urea and for the 4.5-dihydroxy- and 4.5-dialkoxy-ethylene ureas. It can be seen that the ring atoms of the ethylene urea and dihydroxymethyl ethylene urea form practically planar ring systems. This is in agreement with the high electron densities at the nitrogen atoms. The electron pairs at the nitrogen atoms adopt a fixed spatial orientation such that they cannot become coplanar with the electrons of the carbonyl double bond. Therefore, the hydrolysis stability of the N-C- and the C-O-bonds is low. The 4.5-dihydroxy-ethylene urea and the other derivatives listed are non-planar rings. All calculations are based on trans-4.5-dihydroxy(alkoxy) derivatives. The torsion angles are significantly greater. These greater angles are synonymous with lower electron densities at the nitrogen atoms and result in better hydrolysis stability of the N-C- and the C-O-bonds. With an increasing degree of substitution the torsion angles increase. Therefore the hydrolysis stabilities of the N-C- and the C-O-bonds increase too. At the same time the reactivity decreases. Furthermore it must be noted that the dihydroxymethylated and dialkoxymethylated 4.5-dihydroxy(alkoxy)-ethylene ureas contain four reactive groups. Between all of the reactive groups there are highly significant interactions reducing the reactivities too. The change in the torsion angles in combination with the interactions between the reactive groups explains the extraordinary hydrolysis stability of the finished fabrics. The reactivity of the dimethoxymethyl-4.5-dimethoxy-ethylene urea is so low that it is impossible to crosslink cellulose with this compound even in the presence of "hot catalysts".

MODIFICATION OF DIHYDROXYMETHYL-4.5-DIHYDROXY-ETHYLENE UREA

From all chemical and physico-chemical points of view only the hydroxymethylated or alkoxymethylated 4.5-dihydroxy- or 4.5-dialkoxy-ethylene ureas can be used for finishing textiles with high levels of durable press ratings, high levels of crease recovery angles, extremely good hydrolysis stability and a low release of formaldehyde during the finishing process and during the storage of finished fabrics.

$HOCH_2-N$ $N-CH_2OH$ HO OH	$ROCH_2-N$ $N-CH_2OR$ RO OR $R = H, CH_3$	$ROCH_2-N$ $N-CH_2OR$ RO OR Extremely modified
Type A	Type B	Type C

The liberation of formaldehyde can be reduced by decreasing the formaldehyde molar ratio in the hydroxymethylation step, but this often results in lowering the levels of the desired easy-care properties. For reducing the liberation of formaldehyde on finished fabrics and to get a better chlorine fastness, it is possible to use a crosslinker of type B partially etherified with methanol. The chemical nature of the alkoxy group and the position of substitution determine the decrease of formaldehyde liberation. Finishes with a very low release of formaldehyde are based on extremely modified crosslinkers of type C. A very important factor in achieving low levels of formaldehyde liberation is the ratio between crosslinker and catalyst, the type of catalyst and the curing conditions. The most important process variables for an easy-care finish with a pad-dry-cure technique are the concentration of the crosslinking agent, the catalyst ratio, the curing temperature and the curing time.

CROSSLINKER TYPE C
Release of Formaldehyde according to the Japanese Law 112

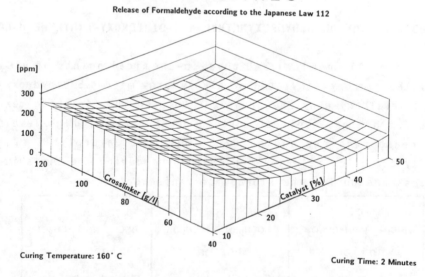

Curing Temperature: 160° C

Curing Time: 2 Minutes

Figure 5. 3 D Contour Plot for the Release of Formaldehyde according to Japanese Test Method Law 112 on finished Fabrics with the Crosslinker Type C.

Using D-optimal statistical design it is possible to obtain all the information about the release of formaldehyde on the fabrics with the various test methods. It can be seen from the 3-D contour plot in Figure 5 that for a constant curing temperature at 160° C and a constant curing time of 2 minutes the crosslinker type C shows, in combination with a special catalyst system formaldehyde levels of less than 100 ppm according to Japanese Test Method 112.

LIBERATION OF FORMALDEHYDE DURING DRYING AND CURING

The liberation of formaldehyde during the drying and curing of impregnated fabrics with glyoxal-based finishing agents leads to the surprising result that the release of formaldehyde during the drying step is several magnitudes higher than during curing.

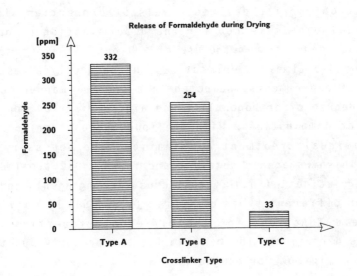

Figure 6. Release of Formaldehyde during Drying of Fabrics with 100 g/l Finishing Agent (45 %) and Catalyst

It can be seen from Figure 6 that a finish with the extremely modified reactant type C reduces the release of formaldehyde liberation during the drying step down to a level of about 30 ppm, with reference to the weight of fabric. That is about only 1/10 in comparison to the common reactant types A and B. The levels of the release of formaldehyde during the curing operation are in the range between 30 and 70 ppm, with reference to the weight of fabric.

OPTIMIZATION OF FINISHING PROCESSES

For every finishing agent in combination with a given type of catalyst there are limited ranges of quantitative ratios, depending on curing temperature and curing time, which result in optimum crosslinking effects with a common optimum of all desired levels of properties, including a minimum of formaldehyde liberation during the finishing process and on the finished fabrics. In all textile finishing procedures the question must be answered under what conditions it is possible to obtain the desired levels of properties simultaneously. The best method for the determination of optimum finishing conditions is the use of a D-optimal experiment design. The D-optimal experiment designs need only a restricted number of experiments, selected in a special manner to reach a high degree of orthogonality and a good distribution of the treatment combinations. With the four process variables only 30 individual treatment combinations are necessary for a given finishing agent, catalyst and fabric. D-Optimal designs were carried out with different crosslinking agents and catalysts on different types of fabric (2,3,4,5). In many cases, the yield functions for the different properties run in opposite directions. The maximum for one property function is often the minimum for another.

Figure 7 illustrates the yield contour profile for the crosslinker type C. At a curing temperature of 160° C and a curing time of only two minutes all desired levels for dry crease recovery angles, tensile strength, loss of abrasion, durable press rating and release of formaldehyde below 100 ppm according to the Japanese Test Method Law 112 are achieved with treatment combinations lying within the hatched area.

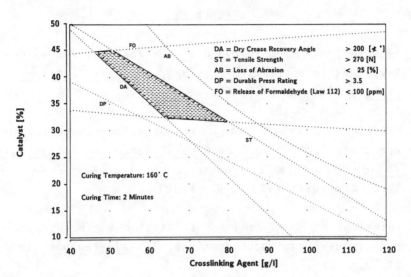

Figure 7. Yield-profile Graph for Finishing Cotton Fabrics with Crosslinker Type C

CONCLUSIONS

The results represent part of an extensive programme of investigations. The interactions between the crosslinking agents, catalysts and fabrics have been studied with kinetically, thermodynamically and molecular modeling methods in combination with modern methods of statistical experiment designs. The results lead to the conclusion that there are special optima for finishing conditions for each crosslinking agent in combination with a given kind and amount of catalyst. Contrary to former thinking, almost all properties of finished fabrics are dependent on significant interactions between the process variables. With this knowledge, it is possible not only to establish agreement between theory and the experimentally observed facts, but also to predict the

limits of finishing effects attainable with a given finishing
agent and catalyst. Furthermore, this knowledges make it
possible to synthesize finishing agents with definite properties.

REFERENCES

1.   Petersen, H., Formaldehyd und neue Konzepte für seine
      Anwendung in der Textilveredlung, Melliand Textilbe-
      richte, 1986, 67, p.p. 656-663.

2.   Petersen, H., Crosslinking with Formaldehyde-containing
      Reactants. In Handbook of Fiber Science and Technology:
      Volume II, ed., M. Lewin and S. B. Sello, Marcel Dekker
      Inc., New York, 1983, p.p. 48-327.

3.   Petersen, H., Anwendung der statistischen Versuchspla-
      nung in der Textilhochveredlung, Textilveredlung, 1977,
      12, p.p. 51-74.

4.   Petersen, H., Optimierung textiler Applikationsverfah-
      ren mit Methoden der statistischen Versuchsplanung am
      Beispiel der Textilhochveredlung mit Dimethylol-4.5-
      dihydroxyethylenharnstoff, Melliand Textilberichte,
      1980, 61, 174-180, 274-281.

5.   Petersen, H. and Pai, P.S., Reagents for Low-Formalde-
      hyde Finishing of Textiles, Textile Research Journal,
      1981, 51, p.p. 282-302.

DISCUSSION

chaired by N. R. Bertoniere (Southern Regional Res. Cent.)

Q: M. Sakamoto (Tokyo Inst. Tech.)
    Which type of the crosslinkers do you sell the most, and
why ?

A: The most important crosslinker types today are still the
dihydroxymethyl-4,5-dihydroxyethylene urea and a partially
with methanol etherified crosslinker. With those types most
of the desired properties are reached. Only the liberation of
formaldehyde during the finishing procedure and afterwards
the release of formaldehyde on finished fabrics for the first
type is in a range of about 500ppm and for the second type in
the range of 200 to 250ppm according to the Japanese test
method Law 112. The latest generation of crosslinking agents
(type C) becomes now more and more important.

Comment: H. Tokura (Nara Women's Univ.)
    I would like to give some comments for your lecture in
terms of human physiology. I suppose it is absolutely essential for clothing wearing comfort that the properties of good
vapour absorbancy in cotton fabrics should not be lost in
treating, modifying and finishing cellulosic fibers chemically and physically, because according to our recent results,
the properties of vapour absorbancy in cotton and wool
fabrics are highly significant in determining the amounts of
human perspiration, the level of clothing microclimate humidity between human skin and undershirts, onset time of sweat-
ing rate, the level of tympanic membrane temperature and so
on. These physiological parameters are known to be closely
related to clothing wearing comfort under various environmental conditions. Thus, human physiology, especially thermal
physiology is deeply influenced by the vapour absorbancy of
cotton fabrics. With these in mind, I would like to emphasize
that the characteristics of good vapour absorbancy in cotton
fabrics should be maintained from the viewpoint of human
physiology when cellulosic fibers are treated, modified and
finished chemically and physically in future research.

NEW CELLULOSE MEMBRANE

KENJI KAMIDE

Fundamental Research Laboratory of Fibers and Fiber-Forming Polymers,
Asahi Chemical Industry Co., Ltd., Takatsuki, Osaka   569, JAPAN

ABSTRACT

Present status of cellulose membrane science and technology is reviewed, with an emphasis on its medical application as hemodialysis and hemo-filtration membranes for artificial kidney(AK), plasma separation and plasma components fractionation, and removal of virus from human blood. History of improvement of cellulose membrane and AK, in which cellulose membrane is installed, is described together with the fundamentally advantageous properties of cellulose membrane, compared with synthetic polymer membranes.   Particle growth theory explaining the underlying mechanism of forming porous polymeric membrane is briefly introduced and the membrane producing condition/ pore characteristics/ membrane performance correlationships are discussed.   Problem of the long-term hemodialysis-associated syndromes and its fine dissolution by improvement of cellulose membrane are exemplified in the following three cases : carpal tunnel syndrome, transient leukopenia and complement activation.

HISTORICAL BACKGROUND

Recently, polymer membrane has been and is attracting an attention as precise media of separating materials.   In fact, some membrane systems were commercialized up to now.   However, the history of polymer membrane for material separation is never short.   TABLE 1 collects milestone scientific discoveries and technological achievements attained past in the field of membrane separation: The first artificial membrane as separation media was cellulose and its derivatives(collodion) and its first application was found in medical treatment.   Repeated and prolonged clinical application of artificial kidney(AK) with cellophane

39

as a therapy of chronical renal failure(clinical hemodialysis) could become possible after an important invention of A-V(arterio-venous fistula) shunt by Quinton et al. in 1960, enabling us to carry out an extracorporeal circulation of blood. Before this invention, AK had been limited only to the treatment of acute renal insufficiency.

TABLE 1
Brief history of cellulose membrane research

Year	Name	Membrane	Objects	Remarks
1748	A.J.A.Nollet	pig's blader	aq.alcohol	osmosis
1846	L.N.Menard	collodion (cellulose nitrate)		surgery
1857	E.Schweitzer	Schweutzer reagent		
1867	M.Traube	copper ferrocyanide (semi-permeable membrane)		dialysis-membrane
1877	W.Pfeffer	copper ferrocyanide formed on unglazed pottery	aq.sucrose	osmotic-pressure
1885	J.H.van't Hoff			van't Hoff's low
1892	C.F.Cross, E.J.Bevan	viscose	fiber	
1908	E.Brandenberger	cellophane	(packing)	from viscose
1914	J.J.Abel, B.B.Turner L.B.Rowentree	collodion	acetylsalicylic-intoxication (in vitro)	albumin loss 1st hemo-dialysis
1933	W.J.Elford	collodion	virus	pore size estimation
1937	W.Thalheimer	cellophane tube	uremia(in vivo)	
1944	W.J.Kolff	cellophane	acute renal failure (lower nephrosis by crash injury)	1st clinical use
1956	S.Leob S.Sourirajan	cellulose-triacetate	saline water	reverse-osmosis
1960	W.Quinton et al.	A-V shunt	chronic renal failure	repeat-use

It was really since 1960s, in which polymer science very rapidly progressed, that polymer membrane became a target of scientific and industrial research: Leob and Sourirajan[1] discovered reverse osmosis in 1956 and then, developed a reverse osmosis membrane based on cellulose triacetate in later 1950s to 1960s. Reverse osmosis is mainly used for pure water production from saline water and 30 years after the discovery, reverse osmosis has a gigantic market amount to $5.3x10^7/year in USA(1986).[2] In 1970s, the hemodialysis type AK became rapidly and widely popularized as a common clinical therapy in USA, Europe and Japan, where cellophane was substituted with regenerated cellulose membrane with pore size 2-3 nm. Study on polymer membrane with pore size of 6-10 nm for precise dialysis has been energetically carried out and ultra-filtration using membrane with pore size 20-100 nm has been commercialized also in 1970s.

Therefore, it seems reasonable to consider that cellulose membrane, as well as other artificial membranes, has been commercialized successively from smaller average pore size to the direction of increasing pore size. This is very understandable because the easiness of production, the controllabilities of pore characteristics and other performances of the membrane decrease quickly with an increase in average pore size, requiring more sophisticated technology.

Cellulosic membranes are now used in various areas. In this article, we briefly introduce the present status of (1)formation mechanism of porous polymeric membrane and (2)application to medical field.

MECHANISM OF FORMING POROUS POLYMERIC MEMBRANE AND ITS PORE CHARACTERIZATION

The performance of porous polymeric membrane, including cellulose membrane, is closely correlated with the pore size and its distribution. In general, the membrane, whose average pore size can be widely and precisely chosen, is most easily produced by dipping the polymer solution into coagulation bath or by evaporating volatile solvents from the solution. Kamide et al. [3] proposed "Particle Growth Theory" to explain the mechanism of forming porous polymeric membrane through the microphase separation. According to their theory, the two-phase separation phenomena and the dynamics of the phase separation play an important role: If the initial polymer concentration of the casting solution v_p^0 is smaller than the critical solution concentration v_p^c, the polymer-rich phase separates as small particles(primary particles) between 10 – 30 nm in diameter. The primary particles amalgamate into larger

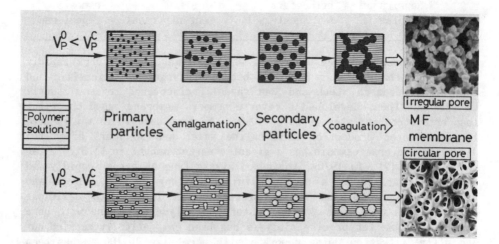

Figure 1. Schematic pore formation process in membrane production.

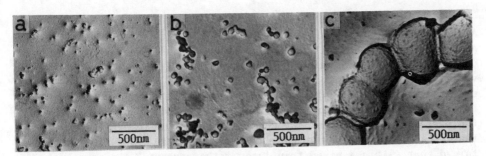

Figure 2. Electron micrographs of a)primary particles, b)growing particles and c)secondary particles in cuprammonium cellulose solution.

secondary particles with diameter($2S_2$) of 50 to 600 nm. The secondary particles subsequently coagulate to form pores. The pore formation mechanism is schematically shown in Fig.1 and the experimental evidence observed during the formation of cupra cellulose membrane by a transmission electron micrographs is shown in Fig.2.

The pore radius distribution N(r), in an imaginary plane within a membrane when $v_p{}^o < v_p{}^c$, is given by :

$$logN(r) = log2N/\{S_2{}^2(1+x)\} + log[r-\{1-(v_p{}^o\rho_s/\rho_p)^{1/3}\}S_2] \\ + [(r/S_2)-1+(v_p{}^o\rho_s/\rho_p)^{1/3}]^2 \times log[x/(1+x)] \qquad (1)$$

with $\qquad x = [R/(1+R)](1/\pi S_2{}^2N) \qquad\qquad\qquad\qquad (2)$

Here, N is the number of pores per unit area and R, the volume ratio of a polymer-lean phase to a polymer-rich phase, ρ_s, the density of the polymer-rich phase, ρ_p, the density of polymer itself. N(r) can be calculated numerically from the radius of the secondary particles S_2 and R. N(r) calculated thus was proved to be in good agreement with that determined by electron microscopic method for cellulose membranes.[3]

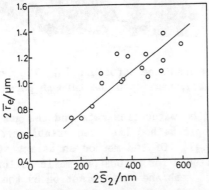

Figure 3. Relationship between mean diameter of secondary particles $2\bar{S}_2$ and mean pore size $2\bar{r}_e$ evaluated by electron microscopic method.

As the theory predicts, mean pore radius and porosity increase with an increase in R or S_2(Fig.3) and R increases with the decrease in v_p^0. Accordingly, the porosity and the mean pore radius increase with the decrease in v_p^0. Cupra cellulose membrane, formed under the condition of $v_p^0 < v_p^c$, has irregular pores(Fig.1, 2) and that formed at $v_p^0 > v_p^c$ has circular pores(Fig.1).[4] The phase separation proceeds usually from the top to the bottom surface and sometimes reverse is true. Although the overall supermolecular structure changes significantly depending on the distance Z from the top surface to a given thin layer, it was ascertained that within a given thin layer with constant Z the supermolecular structure of the layer remained almost uniform. Accordingly, $N(r)$ for each portion of the ultra-thin layer was constant for a given Z.[5]

These experimental facts clearly lead us to the concept that the porous polymeric membrane prepared by the micro-phase separation method should be considered as a composite, in which many hypothetical ultra-thin layers are piled up and when $v_p^0 < v_p^c$, the ultra-thin layer is two-dimensionally composed of many small particles(see, Fig.4).

Then, we come to a fundamental question : What is the pore in a real polymer membrane? Figure 4 shows a schema of a membrane model, in which the membrane is represented as a mixture of secondary polymer particles(filled sphere) and imaginary vacant particles(unfilled sphere).[6] Here, the polymer-lean phase is assumed to be representable by vacant particles, becoming a part of a pore. Our theoretical calculations indicate that for the porosity $P_{re} > 0.4$, the through pores can be most abundant and for $P_{re} < 0.15$, most of pores are the isolated pores.[6]

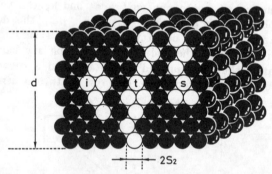

Figure 4. Schema of a membrane structure.[6] The symbols of i,s and t stand for isolated, semi-open and through pores, respectively.

In addition to the water-flow rate and the gas permeation methods, an electron microscopic method has been established for estimating the mean pore radius $\bar{r}_e$.[7] Of the method an aslant sliced thin sectional technique was established for porous polymeric hollow fiber membrane.[8]

Thus, science of membrane pore formation by the micro-phase separation method has been and even now is rapidly progressing. Based on the above scientific advances, we can now design and control, with high accuracy and good reproducibility, the average pore size, the pore size distribution, the porosity and the number of thin-layers consisting a membrane. All of these quantities govern more or less the performance of the membrane. Therefore, this is undoubtedly a key technology in a successful development of new cellulose membrane.

HEMODIALYSIS CELLULOSE MEMBRANE

A very earlier history of artificial kidney(AK) was summarized in the first section (see TABLE 1). AK was widely recognized as a clinical and repeated treatment for chronic renal failure and for this purpose, cellulose membranes regenerated from cuprammonium solution and saponified from cellulose triacetate were commercialized in 1960s. Synthetic polymers[polyacrylonitrile(PAN), polymethylmethacrylate(PMMA), ethylene-vinyl alcohol copolymer(EVA)] membranes for AK were developed in 1970s, but their position in AK market is remaining even at present at comparatively low level. In this sense, we can say that AK was initiated and then advanced along with development of regenerated cellulose membrane, by which the various type of dialysis apparatus was invented : In 1960s a coil type AK was practicized and in 1970s a plate type and a hollow fiber type AKs were commercialized. These three types were soon utilized. But, the coil type, then the plate type decayed in later 1970s – early 1980s. At present, more than 90% of hemodialyzer commercially available is the hollow fiber type. In technical develop-ment of the hollow fiber membrane, cellulose has played always a predominantly leading role: Saponified cellulose acetate(SCA) was first used for cellulose hollow fiber membrane (Cordis-Dow 1967). Then, in 1973 Asahi Chemical Industry developed cupra cellulose hollow fiber membrane. Cordis-Dow commercialized cellulose acetate(CA) hollow fiber membrane in 1977. For these fifteen years, regenerated cellulose membrane has made a prominent improvement in the following points : (1) pin hole, (2) break-down during dialysis, (3) membrane thickness ($100\mu m \rightarrow 7\mu m$), (4) purity (low molecular weight non-cellulosic components), and (5) sterilization method(ethylene oxide, steam, γ rays). World market of artificial hemodialysis apparatus is expanding at rate of 5 % / year and world consumption estimated is 3×10^7 unit/year in 1988. Among them, cupra cellulose membrane occupies 66%, CA membrane, 15% and the re-maining is synthetic polymer membrane. Viscose rayon cellulose membrane, utilized in AK embryo stage, has the weak wet breaking strength, being difficult to produce thin membrane, and now inavailable commercially.

Total number of patients, receiving regular hemodialysis treatment, is estimated in 1986 to be 2.75×10^5[7.2×10^4 each in Japan, USA, and Europe, and 6×10^4 in other countries]. It is also estimated to be above 8×10^4 at the end of 1987 in Japan. Among them, 6.4×10^4 hemodialysis patients are considered to use cellulose and its derivatives membranes. At present, patients receiving hemodialysis for the period longer than 10 years in Japan amount to nearly 1.2×10^4 and the longest record of hemodialysis for a specific patient exceeds really 21 years. This is just the same length as a whole history of AK in clinical use. Such a rapid progress of hemodialysis treatment is strongly supported by the fundamental properties of cellulose membrane, such as (1)chemical stability and safty, (2)wide range of the mean pore size, (3)good balance of solute and water permeability, (4)mechanical strength→thin membrane, (5)easy chemical modification→a variety of performance.

In comparison with synthetic polymer membrane, cellulose membrane has (1)larger removability of low molecular weight nitrogen metabolites such as urea, (2)better balance of material permeability and water-ultrafiltration rate, (3)larger tensile strength in wet state; cupra cellulose membrane is 2.5 - 8 times stronger in wet state than synthetic polymer membranes, (4)higher dimensional stability and good process-ability in module manufacturing.

As number of patients receiving long-term hemodialysis treatment is rapidly increasing, new chronic hemodialysis-associated syndromes appeared : For example, carpal tunnel syndrome ; amyloid protein, deposited in the synovium or the transverse carpal ligament, presses perineural tissue of the median nerve, resulting in acute pain (amyloidosis).[9] This syndrome developes at probability of 70% in patients receiving long-term(≥15 years) hemodialysis treatment. Amyloid responsible for this syndrome is β_2-microglobulin (β_2-MG)[10] with the molecular weight of 11,800 (a molecule is appoximately 4.5nmx2.5nmx2.0nm in size).[11]

Cupra cellulose hollow fiber membrane for hemodialysis(eg., Asahi Medical, AM series) was believed to have the mean pore diameter of 2-3 nm(water-flow rate method), which is too small to allow complete permeation or ultrafiltration of β_2-MG. Recently, new cupra cellulose hollow fiber membrane with larger mean pore size(4-9 nm) was developed by Asahi Medical(AM-2000UP). TABLE 2 shows their typical performance. The AM-2000UP series have evidently larger cut-off point than the AM series. Sieving coefficient S_C of β_2-MG was improved up to 0.4 - 0.6. It is interesting to note that β_2-MG is removed in hemodialysis not by ultrafiltration mechanism, but by diffusion mechanism. By applying this high performance cupra cellulose membrane for long-term hemodialysis, serum β_2-MG level in the patients was proved to decrease effectively.

When cupra cellulose membrane is employed as separation media in extracorporeal circulation of blood, number of leukocyte per unit volume

TABLE 2
Some typical performance of cupra cellulose membrane

Year developed		1984[*1]	1986[*2]	1988
Mean pore diameter /nm		3	4	9
Mass transfer coefficient				
Urea	/cm·sec^{-1}	9.50×10^{-4}	10.0×10^{-4}	9.00×10^{-4}
Vitamine B_{12}	/cm·sec^{-1}	0.85×10^{-4}	1.10×10^{-4}	1.44×10^{-4}
β_2-MG	/cm·sec^{-1}	0.06×10^{-4}	0.21×10^{-4}	0.72×10^{-4}
Sieving coefficient				
β_2-MG		0.04	0.4	0.6

[*1]: Asahi Medical AM series, [*2]: Asahi Medical AM-2000UP series

of blood was found to decrease temporarily down to 20 - 40 % of the
initial value in 15 min after starting the circulation (i.e.,transient
leukopenia). This phenomenon is closely correlated with activation of
complement and some one pointed out that in order to avoid transient
leukopenia synthetic polymer(i.e.,hydrophobic) membrane seems rather
preferable to cupra cellulose membrane. But, this is not true.

Akizawa et al.[12] disclosed that activation of the complement by
cellulose is caused by a direct activation of the second path of
complement due to a contact of the free hydroxyl group on the membrane
with the human blood. And recently a new technique was successfully
developed to suppress the activation of complement: The free OH group on
the membrane surface was masked completely with a cationic polymer[13],
which effectively prevents a direct contact of the OH group with the
blood. The new membrane has a sufficient performance in this
respect(Fig.5). This kind of studies is now rapidly popularizing to
improve the surface properties of cupra cellulose membrane. The copper
content remained in the commercially available cupra cellulose membrane
is kept at very low level (<<10 ppm) and the serum copper of the
patients receiving long-term hemodialysis using cupra cellulose membrane
never exceeds the upper limit of the normal value.[14]

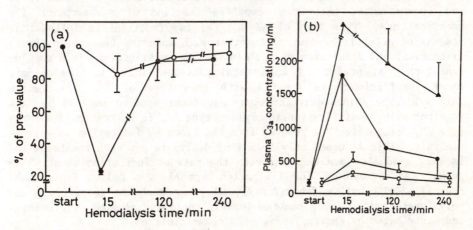

Figure 5. Changes in leukocyte(a) and plasma C_{3a}(b) concentration during
hemodialysis using cellulose membranes[12] : (a) ○, new cellulose; ●,
original cellulose. (b) ○, arterial side and △, venous side of new
cellulose; ●,arterial and ▲,venous side of original cellulose.

ULTRAFILTRATION MEMBRANE

Ultrafiltration(i.e.,hemofiltration)-type artificial kidney

From view point of renal physiological biology, living kidney evolv-
ed from dialysis-type(invertebrate animal)→filtration-type(fish)→
filtration/reabsorption-type(higher vertebral animal).[15] In this way,
the environments govern the pattern of mass transfer of metabolites. In

majority of commercially available AK, the mass transfer is controlled
by diffusion mechanism(dialysis), indicating that the AK used at present
can only be regarded as "artificial kidney for protozoan".

Kamide, Sato and their collaborators carried out since early 1970s,
systematic studies on filtration-type artificial kidney using cellulose
diacetate (CDA, total degree of substitution $\langle\langle F \rangle\rangle$ =2.46) membrane,
having the mean pore radius of 0.01 - several μm.   They[15-18]
clarified that (1)the adequate mean pore diameter of the CA membrane,
suppressing the concentration of serum proteins in the filtrate to 1/10
of that in the fresh blood, is ca. 90 nm, (2)parallel flow is strongly
recommended for use, (3)the components except for proteins in blood are
concentrated by membrane filtration: total protein, x0.05 ; blood urea
nitrogen(BUN), x1.80 ; uremic acid, x1.30 ; Na$^+$, x1.20 ; K$^+$, x1.40 ;
Cl$^-$, x1.75, (4)for removal of water from the patient, this filtration-
type can be applied clinically, but balance of water and metabolites
(urea, creatinine, uric acid) is, as anticipated, not satisfactory.

Filtration-type AK is a kind of artificial glomerulus and CA
membrane corresponds to basement membrane in human kidney.   Therefore,
if artificial tubule could be invented in future, artificial human
kidney in true sense can be realized as an ultimate resolution. Hemofiltration with pre-filtrate or post-filtrate alimentation is practiced to
add conventional, but never complete function of reabsorption for
practical use.   This type of AK has the two functions as follows: 1)
removal of metabolite solutes, together with water by filtration (hemo-
filtration) and 2)infusion of aq. electrolyte solutions as fluid supplementation.   At present, 1 % of total patients suffering from chronic
renal insufficiency is treated with this type of AK.   Filtration
/dialysis-type AK(hemodiafiltration) was first studied in 1974 by Sato
together with hemofiltration/adsorption-type AK. In filtration/dialysis-
type AK, metabolites are removed from the blood by filtration as well as
dialysis.   This AK needs only 5-10 l of dialysate per each treatment and
is very gradually popularizing with the rate of indication about 4% in
Japan.   As a membrane for filtration-type AK, PAN hollow fiber(Asahi
Medical), PMMA hollow fiber(Toray), CA plane membrane (Sartorius) and
polystyrene hollow fiber membrane(manufactured and moduled by Fresenius
and distributed by Kuraray) have been commercialized.

Hemofiltration
In 1977, a possibility of plasma separation by hollow fiber
membrane was first shown by Yamazaki et al..[19] Separation of blood
components such as red and white cells by membrane(i.e.,plasma separation) has been for these 10 years extensively studied as an application
of industrial ultrafiltration technology. A new membrane method was
expected to exchange human plasma more promptly and readily as compared
with conventional centrifuge method.   CA plasma separation membrane is
manufactured and commercialized by Asahi Medical and Cordis-Dow.   The CA
hollow fiber membrane with 0.1 - 0.4 μm in pore size(Asahi Medical) is
used as plasma separation membrane for the treatment of intractable
ascites and utilized in hepatic support system (plasma perfusion detoxi-

cation) or plasma exchange for acute hepatic failure. Furthermore, the CA membrane has been experimentally evaluated and clinically used in renal, hepatic and immuno diseases, not only to remove toxic macromolecular substances, but also to supply necessary nutrients.[20] Besides cellulose derivatives, PMMA(Toray), polyvinylalcohol(Kuraray), polyethylene(Mitubishi rayon) and polypropylene(Travenol, Gambro) are non-cellulosic plasma separation membranes commercialized up to now.

As a result of development of plasma separation membrane, double-filtration plasmapheresis, in which selective removal of large molecular weight, disease-related materials is carried out using the membrane, is now established as new therapy. In this treatment, plasma separation membrane with mean pore size of 0.1 - 0.4 μm is called as primary membrane and the membrane with mean pore size of 10 - 80 nm for fractionation of plasma components, as secondary membrane(Fig.6).

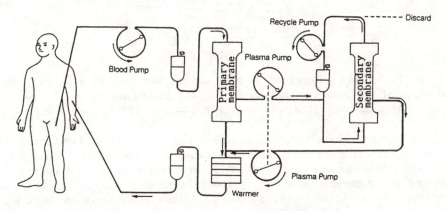

Figure 6. Schematic representation of double-filtration plasmapheresis.

Removal of virus from human blood by filtration method

Very recently, acquired immune deficiency syndrome(AIDS) virus (Human Immunodeficiency virus,HIV) and hepatitis B virus(HBV) found in human blood, has been attracting an enormously keen social attention, and establishment of a method for the effective removal of these viruses is urgently called for as a clinical therapy or in fractionated plasma derivatives manufacturing. This is, in a scientific words, a method of removal of the viruses from coexisting serum proteins. Compared with a conventional membrane separation technology applied for other fields, including industry and AK, any possible method designed for the removal of virus from aq. serum protein mixtures(human blood) should have (a) extremely high rejection rate(usually, ≥99.99%) of virus and (b)very large permeability of serum proteins(for example, ≥95% for albumin).

Figure 7 shows the adsorption of albumin from saline solution onto various polymer membranes. Here, BMM is the porous cupra cellulose membrane invented by Asahi Chemical Industry. Obviously, cellulose has the least tendency of protein adsorption, and is, we believe,

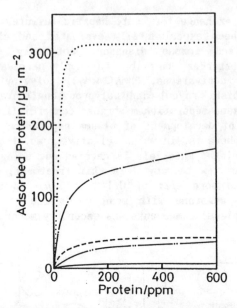

Figure 7. Protein adsorption of various polymer membranes at 25℃: The solution of bovine serum albumin in 0.9 wt% saline water; BMM, ——— ;PMMA, —··— ;CDA, ······· ;PAN, —·— ;PTFE, ············ .

only a polymeric material, which allows the practical application. The adsorbed proteins on cellulose membrane(i.e., hydrophilic membrane), although their amounts are very small, are easily desorbed from it, but hydrophobic membrane including synthetic membrane, adsorbs strongly and irreversibly proteins. The porous cellulose membrane with mean pore diameter ranging from 20-100 nm and having sharp pore radius distribution, will enable us to remove exclusively disease viruses contaminated in useful proteins(plasma) and concurrently to recover these proteins with high rate. Asahi Chemical Industry succeeded to develope the porous cupra cellulose hollow fiber membrane BMM for this purpose and it is now under test pilot scale production: We can now prepare the membrane with any mean pore diameter(water-flow rate method), arbitrarily chosen between 10 -100 nm. The composition of filtrate, separated by BMM with mean pore diameter larger than 20 nm, is very similar to that of original untreated plasma. TABLE 3 shows the retention ratio of the activity of fresh human plasma before and after filtration by BMM with various mean pore diameter.[21] Note that in plasma filtration through BMM, the activity of 8th coagulating factor (F Ⅷ) is highly maintained and interestingly, the complement is not activated, indicating that activation of complement, often observed in clinical hemodialysis using cellulose membrane, is not due to intrinsic nature of cellulose.

The rejection coefficients of various disease and non-disease viruses, contaminated in culture media, human plasma, bovine serum and other, by BMM were determined systematically. When BMM with mean pore diameter

TABLE 3
Retension ratio of activity of fresh human plasma before and after
filtration by BMM[21]

Pore size	Fig Tb Time/%	FVⅧ APTT/%	FⅨ APTT/%	ATⅢ *1/%	C consumpt. CH$_{50}$ /%	FN ELISA/%
NO	100	100	100	100	100	100
50 nm	106.0	86.4	91.2	102.0	94.4	112.0
20 nm	90.3	99.4	89.9	67.6	91.7	98.6
10 nm	15.8	2.4	21.0	16.7	41.7	62.9

*1:Chromogenic method

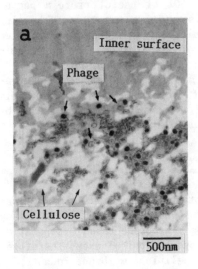
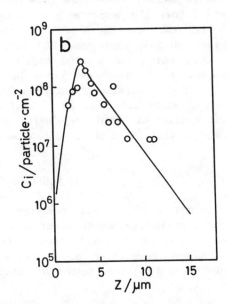

Figure 8. Transmission electron micrograph of T4 phage captured in BMM
with mean pore size of about 40.6nm(a) and its concentration
distribution in the BMM(membrane thickness,28 μm; porosity,50.4%) after
filtration(0.5 ml/cm²; pressure difference,200mmHg)(b): C_i,number of
phage per cm² ; Z, distance from inner surface of hollow fiber membrane.

smaller than 105 nm was used for HIV, HIV concentration, determined by
plaque method, was confirmed to be far below the detectable limit.
The movement of virus in porous membrane during filtration is directly
observable by electron microscopy(Fig. 8a).   Figure 8b shows the
concentration distribution of Escherichia coli phage T4 within the BMM,
which can be reasonably interpreted by the probability process. The
filtration of virus through BMM can be regarded as multi-stage(or multi-
repeated) filtration, resulting in an extremely high separation
efficiency. From Figure 8b, the probability of capturing a given virus
in the i-th layer γ_i was calculated and by using this value, the virus
rejection coefficient of the membrane ϕ [=log(A_i/P_N), A_i, the total
number of virus applied to a membrane, P_N, the total number of the virus

flowed out from the membrane] was estimated to be 5.25, which agrees with the experimental value estimated by the plaque method (4.02). It is now evident that even if γ_1 in each layer is low(0.1, in this case), sufficient high ϕ can be attained if the number of layers exceeds almost 100. This is a physical explanation why BMM has such a high performance. The ϕ is a function of the total amount of the filtrate and/or the filtration time and there exists a critical amount of filtrate, under which a given level of ϕ is absolutely maintained. However, this does not deny a high ϕ of BMM for the practical purpose. In the case of exclusion of virus from serum proteins by using membrane, it is needed that the membrane has more than 50% of useful protein permeability and more than 4 of ϕ in comparison with a conventional heating method actually employed. And this requirement is now fully satisfied by BMM, which has a good possibility of application to Cohn's cold ethanol precipitation procedure for serum protein production.

Other fruitfully expected usages of BMM are: (1)a method of prevention of virus infection by filtration of fresh human blood immediately before clinical blood-transfusion and (2)a therapy of virus disease by membrane filtration of extracorporeally circulated blood of patients.

CONCLUDING REMARKS

Cellulose membrane has numerous advantages over synthetic polymer membranes and cellulose membrane-based separation technology has many merits compared with non-cellulosic membrane-based separation technology, especially for medical and biotechnological uses. Therefore, nobody can doubt the prosperous future of cellulose membrane industry.

ACKNOWLEDGEMENTS

The auther would like to express his sincere gratitude to Mr. Hideki Iijima, Senior Scientist of the laboratory for his invaluable cooperation in preparing the manuscript.

REFERENCES

1. Leob,S., The leob-sourirajan membrane: How it came about. In Synthetic Membranes: Desalination, Am. Chem. Soc. Sympo. Ser., No.153, 1981,pp.1-9.
2. Crull,A.W., Business Oppotunity Report, P-041U, Membranes Separations Markets and Technologies, Business Communications Co., Inc., Stamford, 1986, p.xiv.
3. Kamide,K. and Manabe,S., Role of Microphase Separation Phenomena in the Formation of Porous Polymeric Membrane. In Material Science of Synthetic Membranes, ed.,D.R.Lloyd, Am.Chem.Soc.Symp.Ser., No.269, 1985,pp197-228.
4. Kamide,K., Iijima,H. and Iwata,M., Effect of Preparing Conditions on Supermolecular Structure of Cellulose Micro-Porous Membrane. Polymer

Preprints,Japan(English Edition), 1988,37,Nos.5-10,E419.
5. Kamide,K. ,Kamata,Y, Iijima,H. and Manabe,S., Some Morphological Characteristics of Porous Polymeric Membranes Prepared by "Micro-Phase Separation Method". Polym.J.,1987,19,391-404.
6. Manabe,S., Iijima,H. and Kamide,K., Probabilities of Finding Isolated, Semi-Open, and Through Pores in Porous Polymeric Membrane Prepared by Micro-Phase Separation Method. Polym.J.,1988,20,307-319.
7. Manabe,S., Shigemoto,Y. and Kamide,K., Determination of Pore Radius Distribution of Porous Polymeric Membranes by Electron Microscopic Method. Polym.J., 1985,17,775-785.
8. Kamide,K., Nakamura,S., Akedo,T. and Manabe,S., An Electromicrographical Method for Evaluating Layer Structure of Polymeric Hollow Fiber Membrane. Polym.J., to be submitted.
9. Vandenbroucke,J.M., Huaux,J.P., Guillaume,Th., Noel H., Maldague,B. and C van Ypersele de Strihou, Capsular Synovial and Bone Amyloidosis :Complications of Long-Term Haemodialysis. Proc.EDTA-ERA,1985,22,136-138.
10. Gegyo,F., Yamada,T., Odani,S., Nakagawa,Y., Arakawa,M., Kunitomo,T., Kataoka,H., Suzuki,M., Hirasawa,Y., Shirahama,T., Cohen,A.S. and Schmid,K., A New Form of Amyloid Protein Associated with Chronic Hemodialysis Was Identified as β_2-Microglobulin. Biochem.Biophys. Res.Commun.,1985,129,701-706.
11. Becker,J.W. and Reeke,G.N.Jr., Three-Dimentional Structure of β_2-Microglobulin. Proc.Natl.Acad.Sci.USA, 1985,82,4225-4229.
12. Akizawa,T., Kitaoka,T., Koshikawa,S., Watanabe,T., Imamura,K., Turumi,T., Suwa,Y. and Eiga,S., Development of a Regenerated Cellulose Non-Complement Activating Membrane for Hemodialysis. Trans.Am.Soc. Artif.Intern.Organs, 1986,32,76-80.
13. Corretge,E., Kishida,A., Konishi,H. and Ikada,Y., Grafting of Poly (ethYlene Glycol) on Cellulose Surfaces and the Subsequent Decrease of the Complement Activation.In Polymers in MedicineⅢ,ed.C. Migliaresi et al., Elsevier Science Publishers B.V.,Amsterdam,1988, pp.61-72.
14. Ohnishi,M., Serum Copper Changes in Patients on Long-Term Hemdialysis. J.Osaka City Med.Cent., 1986,35,189-205.
15. Kamide,K., Manabe,S., Sato,K., Hamada,K. and Miyazaki,S., Marginal Ability of Filtration Type Artificial Kidney. Jpn.J.Artif.Organs, 1975, 4, 220-225.
16. Sato,K., Improvement of Filtration-Type Artificial Kidney-Filtration /Dialysis-Type Artificial Kidney and Filtration/Adsorption-Type Artificial Kidney-. Jpn.J.Artif.Organs,1974,3,422-429.
17. Kamide,K., Manabe,S., Sato,K., Hamada,K., Okunishi,H. and Miyazaki, S., A Method for Evaluating the Ultrafiltration Value of Artificial Kidney in vivo by Using the Date in vitro. Jpn.J.Artif.Organs, 1975, 4,171-174.
18. Kamide,K., Manabe,S., Hamada,K., Okunishi,H. and Miyazaki,S., Mechanism of Permselectivity of Porous Membrane Used in Filtration Type Artificial Kidney. Jpn.J.Artif.Organs,1976,5(Suppl)139-142.
19. Yamazaki,Z., Fujimori,Y., Sanjo,K., Kojima,Y., Sugiura,M., Wada,T., Inoue,N., Sakai,T., Oda,T., Kominami,N., Fujisaki,U. and Kataoka,K., New Artificial Liver Support System(Plasma Purfusion Detoxification) for Hepatic Coma. Artif.Organs,1978,2(Suppl),273-276.
20. Yamazaki,Z., Inoue,N., Fujimori,Y., Takahara,T., Oda,T., Ide,K., Kataoka,K. and Fujisaki,Y., Biocompatibility of Plasma Separation of an Improvement Cellulose Acetate Hollow Fiber. In Plasma Exchange Plasmapheresis-Plasma Separation,ed. H.G.Sieberth, Sehattauer, Stuttgart-New York,1980,pp.45-51.
21. Okuyama,K., Fukushima,Y., Miura,Y., Honma,R., Manabe,S., Ishikawa, G., Satani,M., Komuro,K., Function Analysis of Human Plasma Proteins Passed through the Micro-Porous Regenerated Cellulose Membrane(BMM) Hollow Fiber, Japanese J. Clinical Hematology,1988,29,662-665.

DISCUSSION

chaired by W. Pusch (Max-Planck-Inst. Biophysik)

Q: C. Schuerch (SUNY College Enviro. Sci. Forest.)
Has there been interest in cellulose membranes for gas separations ?

A: Yes, there are some attempts at utilization of cellulose membrane in gas separation. But, unfortunately cellulose does not seem, at least for me, to have unquestionable advantage.

Q: R. Atalla (Inst. Paper Chem. USA)
(1) Do you believe the special properties of cellulose, with respect to the balance between hydrophilic & hydrophobic character are necessary for its performance ?
(2) Do you believe a β-1,4 linked mannan might be equally effective ?

A: (1) Yes, this balance should be a necessary condition, but not a sufficient condition.
(2) I do not know.

FUNDAMENTAL AND INDUSTRIAL ASPECTS OF STARCH GELATION

Paul COLONNA
Centre de Recherches Agro-alimentaires
Institut National de la Recherche Agronomique
B.P.527. rue de la Géraudière. 44026 Nantes Cedex 03. France

ABSTRACT

When processed by heat treatment in the presence of water, starch granules undergo gelatinization. By cooling gelation occurs in two steps: (i) a phase separation which produces a polymer-rich phase and (ii) a crystallisation within this phase. All crystalline phase is of B-type.
An interconnected three-dimensional network is formed with interchain association over length of ≈ 130Å. Water is entrapped in these three-dimensional networks, in which macromolecular probes (enzymes) can diffuse.
This macromolecular reorganisation has major effects on texture, and nutritional value of foods containing starch. Technological control of texture is possible by the dispersion level of starch granules, the amylose and water content, the storage conditions. Mung bean noodles, which are made by gelation of mung bean starch, give low plasma glucose and insulin responses, in contrast to pasted starches.

INTRODUCTION

Starch is the major polysaccharide reserve material of photosynthetic tissues and of many types of storage organs such as seeds, swollen stems and roots. In chemical terms, starch is a mixture of amylose and amylopectin localized within starch granules[1]. Amylose is a linear macromolecule made of α-D-glucopyranose units linked through α-(1->4) linkages; amylopectin is a large branched macromolecule based on short α-(1->4) glucan chains which are joined together through α-(1->6) branch-points. On average, there is one branch-point for every 20-25 main-chain residues. The glucan chains are thought to have a bimodal distribution of chain length in which the most abundant species have degrees of polymerisation (d.p.) of 50-60 and 15-20. The molar ratio of short to long chains varies between 3:1 and 12:1, depending on the botanical source. These short chains of amylopectin are thought to be arranged in crystallites that would be responsible for the crystallinity of the native granule.

Both are organised in cold-water-insoluble granules which are semicrystalline and lack any specific functional properties. The present review deals with gelation, which results from the physical modifications of starch.

PHYSICOCHEMICAL MECHANISMS

When processed by heat treatment in the presence of water, starch granules undergo gelatinization, leading to a mixture of solubilised macromolecules, mainly amylose, and swollen residues from starch granules. By cooling gelation occurs in two steps[2]: (i) a phase separation which produces polymer-rich and polymer-deficient regions and (ii) a crystallisation within this polymer-rich region. If the macromolecule concentration is sufficiently high, the polymer-rich regions form an interconnected gel network. For amylose gelation occurs only above C^* (semi-dilute region), which coresponds to a concentration of ≈ 1.5%. For amylopectin, chain entanglements begin at 0.9%; however gelation starts to occur at concentration of >10%, where the amylopectin chains are heavily entangled.

An interconnected three-dimensional network is formed with interchain association over length of ≈ 130Å. These crystallites are composed of long constitutive chains (d.p. 35-40), are acid resistant and exhibit a high fusion temperature (≈ 153°C) with an enthalpy change (9.4 mJ.mg^{-1}). Amylopectin when present in concentrated solution (>10% w/w) can form gels which melt ≈ 55°C (enthalpy transition 3-15 mJ.mg^{-1}). These crystallites are thin (composed mainly of linear chains of d.p. 15), acid labile. Exterior amylopectin chains crystallise during gelation.

The X-ray diffraction pattern associated with gelation is the B-type, as in tuber starches. Recently the three-dimensional structure of B-starch has been revisited [3]. The crystals are based on a unit cell (a = b = 1.85nm; c = 1.04nm), containing 12 glucose residues located in two left-handed, parallel-stranded double helices packed in a parallel register. The valence angles at the glycosidic linkage are φ ≈ 84° and φ ≈ 144°. 36 water molecules are located between these helices, with no apparent sign of a disorder. They are on six fold helices, following the crystalline symmetry: half of them are connected by hydrogen bonds to an amylosic chain and to other water molecules, and the other half is only connected to water molecules.

Water is entrapped in these three-dimensional networks [4], in which macromolecular probes (enzymes) can diffuse. Diffusion coefficient (D) of BSA in amylose and amylopectin gelled networks decreases with increasing polysaccharide concentration. D decreased from 5.5 x 10^{-11} m^2.s^{-1} to 1.5 x 10^{-11} m^2.s^{-1} over the concentration range 5-15 %w/w of gelling polymer. Accessibility have been studied by diffusion until equilibrium: probe size and macromolecule concentration are the two main parameters. For probes where $R_H<10Å$, the volume of solvent trapped within the network is completely accessible to the probe. Accessibility is a reasonably continuous function of hydrodynamic radius.

Modelling has been carried out when considering gel network as a collection of immobilized, randomly spaced, rigid rods. The particles diffuse through the network by taking directionally random steps: if collision with the network occurs, the random step is not completed and the diffusion retarded. More recently, hydrodynamic screening has been thought to be a more appropriate description of the physical process responsible for the retardation of diffusion. The relationship between the average mesh size, ξ, of the semi-dilute polymer solution and the hydrodynamic radius, R_H, of the probe particle is important. When $R_H>>\xi$, the matrix appears as a continuum, and the Stokes-Einstein relationship may be used to describe the diffusion process with the viscosity term being the macroscopic viscosity of the polymer solution.

FUNCTIONAL PROPERTIES

This macromolecular reorganisation has major effects on texture, and nutritional value of foods containing starch. Amylopectin gels behave as Hookean solids at strains <0.1, amylose gels at strains < 0.2. In the concentration range 10-25% after storage for 6 weeks at 1°C, there is a linear relationship between modulus and concentration, in contrast to amylose gels, where a 7th power dependence of modulus on concentration is observed. Technological control of texture is possible by the dispersion level of starch granules, the amylose and water contents, the storage conditions. Whatever macro-

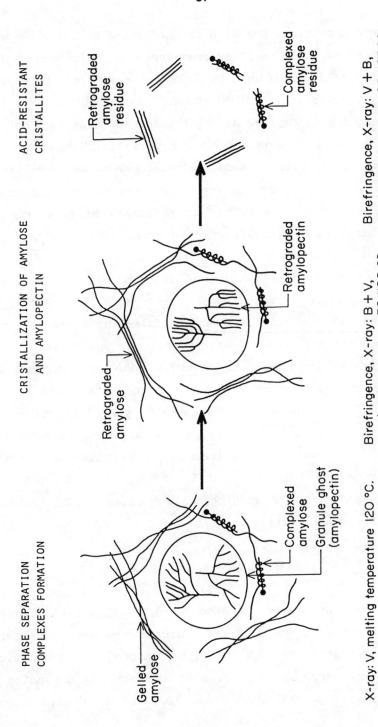

PHASE SEPARATION
COMPLEXES FORMATION

CRISTALLIZATION OF AMYLOSE
AND AMYLOPECTIN

ACID-RESISTANT
CRISTALLITES

Gelled amylose

Complexed amylose

Granule ghost (amylopectin)

Retrograded amylose

Retrograded amylopectin

Retrograded amylose residue

Complexed amylose residue

X-ray: V, melting temperature 120 °C.

Birefringence, X-ray: B + V, melting temperature 50/120 °C.

Birefringence, X-ray: V + B, melting temperature 106/125 °C, DP, 40 – 45.

FIGURE 1: STRUCTURAL FEATURES OF GEL OBTAINED FROM PASTED MAIZE STARCH.

molecular composition, gelation is more rapid at lower temperature. After pasting (Fig.1) or drum-drying of maïze starch [5], amylose crystallises independently from amylopectin by complexes formation and retrogradation. This phenomenon is explained by the leaching of amylose during the thermal treatment and the subsequent phase separation, then crystallisation within the amylose volume fraction. These crystallites, composed of linear chains (d.p. 35-40), are acid-resistant and exhibit a high fusion temperature (above 100°C). In contrast, after extrusion-cooking, amylose and amylopectin co-crystrallize in the same manner as pure amylopectin.

Mungo starch vermicelli and rice flours noodles [6] are based upon these starch networks. In both types of foods, an amylopectin-based structure shows the following characteristics – B-type X-ray diffraction pattern, melting endotherm at 50°C, high susceptibility towards acid attack. Furthermore, amylose-based structures are present mainly in the complexed (V-type diffraction pattern) and retrograded (B-type diffraction pattern) forms for rice flours noodles and mungo starch vermicelli respectively. These crystallites are more resistant towards acid hydrolysis and cooking (melting temperature 82 and 119°C). Cooking behaviour of these glutenless noodles can be explained by these amylose networks.

NUTRITIONAL PROPERTIES

Recent studies have compared the digestion kinetics of starch foods by humans. Raw starches are partly digested whereas all transformed starches are completely. Dietary carbohydrates are classified as "rapid" and "slow", depending on their effect on postprandial blood glucose and insulin responses. "Rapid" carbohydrates are absorbed at a high rate, producing a rapid and high blood glucose rise and a correspondly high insulin response. The duration of the blood glucose elevation is short. In contrast "slow" carbohydrates give a low blood glucose peak with a longer duration and a less pronounced insulin response. "Slow" carbohydrates are regarded essential in the diabetic diet. Furthermore they are also useful for healthy men as insulin may be of importance in vascular disease and hypertension. Mung bean noodles, which are made by gelation of mung bean starch, give low plasma glucose and insulin responses, in contrast to pasted starches such as in bread or extruded flours [6]. The metabolic responses and in vitro digestibility of starch gels are inversely correlated to the amylose content of the considered starch (manihot 17%; wheat 27%; smooth pea 35%). The presence of this starch network owing to high amylose content and specific processing is the structural cause for the low in vitro hydrolysis rate and low glucose and insulin plasma responses encountered. This feature should be explained by the slow diffusion of α-amylase inside the particle gel, macromolecular network being the limiting factor. When hydrolysis and adsorption rates are high, the glycemic response is intense, followed by the same phenomenon for insulin. Nutritionists are now looking for starchy foods able to give low glucose and insulin responses, in order to avoid these diet-related diseases. Starch gels would be a solution to this problem.

CONCLUSION

Our understanding of gelation is growing rapidly due to the use of multidisciplinary techniques for studying the physico-chemical basis. Attempts to gain an unambiguous understanding have not been yet completely successful, specially for the crystallite arrangements and network topology. Industrially if appropriate process are carried out, gelation may be conducted for creating new foods, with designed functional and functional properties.

ACKNOWLEDGEMENTS

The author is deeply indebted to his colleagues from the Institut National de la Recherche Agronomique for helpful discussions.

REFERENCES

1. GUILBOT, A. and MERCIER, C., Starch In The polysaccharides, vol.3, ed., G.O. Aspinall, Academic Press, Inc., New-York, 1985, pp. 209-282.

2. RING, S.G., COLONNA, P., I'ANSON, K.J., KALICHEVSKY, M.T., MILES, M.J., MORRIS, V.J. and ORFORD, P.D., The gelation and crystallization of amylopectin. Carbohydr. Res., 1987, **162**, 277-293.

3. IMBERTY, A. and PEREZ, S., A revisit to the three-dimensional structure of B-type starch. Biopolymers, 1988, **27**, 308-325.

4. LELOUP, V., COLONNA, P. and RING, S. Diffusion of a globular protein in amylose and amylopectin gels. Food Hydrocolloids, 1987, **5/6**, 465-469.

5. MESTRES, C., COLONNA, P. and BULEON, A. Gelation and crystallization of maize starch after pasting, drum-drying or extrusion-cooking. J. Cereal Sci., 1988, **7**, 123-134.

6. MESTRES, C., COLONNA, P. and BULEON, A. Characteristics of starch networks within rice flour noodles and mungo starch vermicelli. J. Food Sci., 1988, accepted for publication

7. BORNET, F.R.J., FONTVIEILLE, A-M., RIZKALLA, S., COLONNA, P., MERCIER, C. and SLAMA, G. Insulin and glycemic responses from different native starches according to food processing in normal subjects. Correlation to their in vitro alpha-amylase hydrolysis. 1988, Am.J. Clin. Nutr., 1988, accepted for publication.

DISCUSSION

chaired by K. Nishinari (Nat. Food Res. Inst.)

Q: P. Zugenmaier (Tech. Univ. Clausthal)
    You have presented the crystalline structure of β-amylose
to consist of parallel running chains. In the original X-ray
fiber diffraction study of Wu and Sarko, they started from
anti-parallel running chains of amylose triacetate, deacetylated the compound and obtained a fiber of crystalline β-amy-
lose. How do you explain the chain reversed from antiparalled
to parallel running chains ?

A: Presented results are from Buleon and Perez. They result
from computing, based on experimental results, including
those of Wu and Sarko. Therefore, they have to be considered
as a new model for crystalline β-type structure. This proposed model presents the advantage of being more in agreement
with all biosynthesis mechanisms.

Q: K. Nishinari (Nat. Food Res. Inst. Japan)
    (1) The effect of molecular weight on the gelation kinet-
ics and the concentration dependence of shear modulus ?
    (2) The similarities and differences between cellulose
and starch in acid hydrolysis ?

A: Dr. Gidley of Unilever is publishing a paper in "Macromolecules", where the molecular weight dependence of gelation
has been extensively studied. In previous publications,
molecular weight is rarely taken into account as this parameter can be controlled only by the synthesis.
    Starch is much sensitive to acid hydrolysis than cellu-
lose:this is due to its solubility in acidic water solutions.

Comment: A. Blažej (Slovak Tech. Univ.)
    Starch differs from cellulose from point of view of solu-
bility, because cellulose structure is rigid and hydrophil OH
groups are blocked by intra- and intermolecular hydrogen
bonds and a solvent for cellulose must be more or less non
polar and must be able to split hydrogen bond. Cellulose has
rigid structure. On the contrary, starch structure is more
flexible, hydroxyl groups in starch are bonded by water and
of more polar structure and thus better soluble in water.

Comment: R. H. Atalla (Inst. Paper Chem. USA)
    With respect to the problem of relative solubility of
starch and cellulose, there is an analogous situation among
the inositols which are the cyclohexane hexols. Scyllo-
inositol, with all hydroxyl groups equatorial, is less than
1% soluble in water. Myo-inositol, which was only one hydroxyl group axial, is soluble in excess of 30%. Neo-inositol,
which has two axial hydroxyl groups across the ring from each
other, is again less than 1% soluble. The two less soluble
molecules have a high percentage of hydrophobic surface, in
spite of the large number of hydroxyl groups. In the same way

cellulose is like a ribbon with hydrophobic surfaces and hydrophilic edges. However, its stiffness also is a factor in its insolubility.

Q: J. F. Kennedy (Univ. Birmingham)
By what methods are you able to measure the amylose/amylopectin ratios in starches and what starches did you use to provide the range of amylose/amylopectin ratios you showed in your last slide ?

A: Experiments reported here have been carried out, using purified amylose and amylopectin from potato and waxy maize, respectively. Purity has been checked by iodine complexation using amperometric method. When working on industrially processed starches, maize starches with varying amounts of amylose were selected. It should be noticed that molecular degradation may lead to an amylopectin over-estimation. In such cases, amylopectin content is measured by the amount of DP 15 and 45 chains, after enzymatic debranching experiments.

Q: J. F. Kennedy (Univ. Birmingham)
Please give information on the molecular size of amylose chains which undergo retrogradation and state how big the molecules must be to show this phenomenon ?

A: The minimum size of amylose chains for gelation instead of retrogradation is around 150. Pfannemueller and more recently Gidley have observed this, using monodisperse amylose chains made by elongation using phosphorylase. Below this critical value, phase separation is followed immediately by crystallization, resulting in the formation of small crystallites.

Q: S. U. Hossain (Kimberly-Clark Co.)
Certain starch derivatives such as cationic starch are substantive to cellulose fibers (wood pulp). Can you speculate on the type of bond between cationic starch and pulp fibers ?

A: Cationic starches are linked to cellulose fibers by electrostatic bonds, involving the charge of cationic starches and the zeta potential of cellulosic fibers. They are strong enough for enabling paper flocculation. However they may become weak during paper recycling. In such cases, pollution would arise, as in European factories.

CHITIN AND ITS NEW FUNCTIONALIZATION.

SEIICHI TOKURA
Department of Polymer Science, Faculty of Science,
Hokkaido University, Sapporo 060, Japan

ABSTRACT

The biodegradability of chitin was enhanced by the destruction of its rigid crystalline structure by the regeneration or chemical modifications. The activation of mouse peritoneal macrophages was observed for fairly long stage on the partially deacetylated chitins and the maximum degree of activation was shown by 70% deacetylated chitin(DAC-70). The sustained release of drug was achieved successively by the hypodermic injection of drug bound 6-0-carboxymethyl-chitin(6-0-CM-chitin) saline solution into rabbit due to the highly biodegradable property of 6-0-CM-chitin. But the production of hapten specific antibody was induced by the hypodermic injection of hapten bound 6-0-CM-chitin with Freund's complete adjuvant. A non toxic chitin heparinoid was prepared by the 3,6-0- and N-sulfations of 70% deacetylated 6-0-CM-chitin which showed 1/3 of anticoagulant activity as that of heparin and induced little hypertension of mouse organs for two weeks after the intravenous injection.

INTRODUCTION

Chitin, a natural abundant mucopolysaccharide, is known to be an analogue of cellulose on the chemical structure except an acetamide group at C-2 position of Glucose residue(2-deoxy-2-acetamide-cellulose) as shown in Scheme 1. The reactivity and solubility of chitin are shown to be remarkably poorer than those of cellulose owing to the high crystallinity supported by hydrogen bonds mainly through acetamide group. But one of advantages of chitin over cellulose is the biodegradability in animal body, although it is very slight degree. There are a number of investigations on the application of chitin as a biomedical material because of its biodegradability in animal body. However, it is so hard to regenerate chitin owing to poor solubility that chemical modification has been required to enhance its solubility. The chemical modification of chitin has to proceed under drastic condition due to the high crystallinity, although the milder reaction condition is required to protect acetamide and glycoside linkages. Thus several reaction conditions have been proposed to modify chitin molecule and some biological properties of these chitin derivatives were investigated fundamentally to apply for the biomedical material.

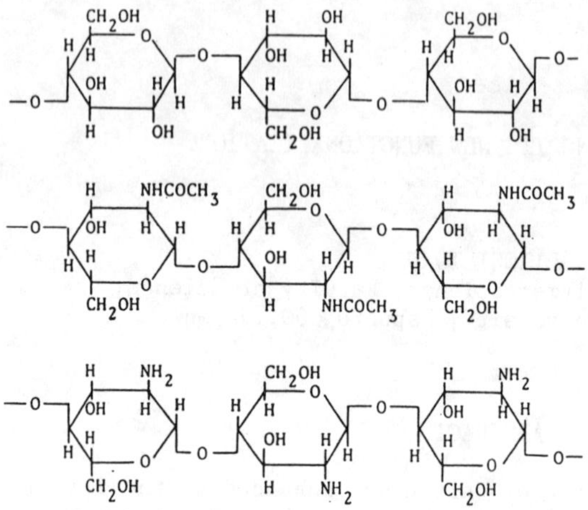

Cellulose

Chitin

Chitosan

Scheme 1. Chemical structures of chitin, chitosan and cellulose.

MATERIALS AND METHODS

Chitin and Chitosans

Chitin was prepared from Queen Crab shells according to the method of Hackman(1) and powdered to 60–120 mesh before use. Deacetylation of chitin was achieved in the mixture of 40% NaOH(w/v) and i-propanol(1:1 v/v) at refluxing temperature for 5 hours to prepare 30–40% deacetylated chitins(DAC 30–40). Then DAC 30–40 was retreated under the same condition for various reaction time following to the rinse with distilled water. The degree of deacetylation was estimated both by elemental analysis and IR spectrum(2).

Chemical modifications of chitin

O-Acetylations of chitin were achieved under the heterogeneous condition using acetic anhydride-perchloric acid system(3). O-Alkylations including carboxymethylation were performed under the condition reported previously(4). Sulfations of chitin and its derivatives were also achieved according to the method reported previously(5). The degree of substitution was estimated mainly by elemental analysis. Phosphorylation of chitin was performed under the condition of previous report applying phosphorous pentoxide-methanesulfonic acid mixture (6). The abbreviations and chemical structures of chitin derivatives are listed in Scheme 2.

Assay for macrophage activation

(i) Harvest of peritoneal macrophages: Mice were injected with each polysaccharide intraperitoneally as a solution or a suspension in 0.5mL of PBS. Peritoneal cells were harvested 3 to 15 days after the injection by lavage with 10mL of HBSS(Hank's balanced salt solution) containing 10 units of heparin/mL. Cells were collected by centrifugation at 200g and resuspended in RPMI-FCS(Nissui Seiyaku RPMI 1640 medium and 10% heat-inactivated fetal calf serum), and plated in 96-well micro tissue culture plate according to the procedure described previously(7). The uniformly adherent cells were incubated for 2 hours at 37°C and then non-adherent cells were removed by extensive rinse with RPMI-FCS.

(ii) Assays for cytolytic activity, for cytostatic activity and hydrogen peroxide release by macrophages: These assays were made according to previous methods as shown in Scheme 3.

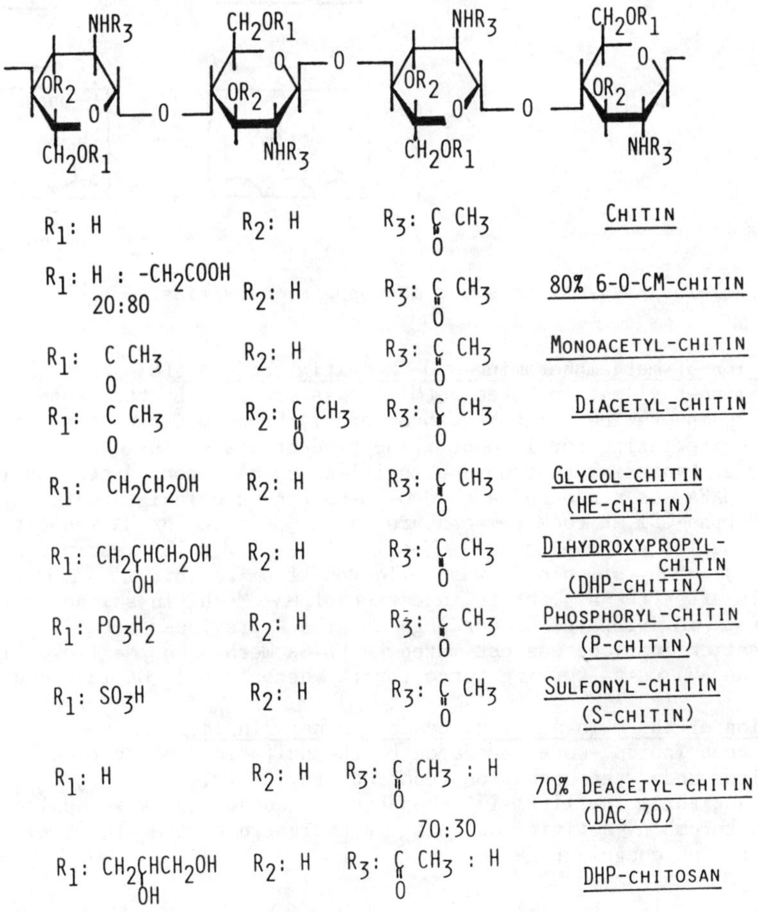

R_1	R_2	R_3	
H	H	$\overset{}{\underset{O}{C}}CH_3$	CHITIN
H : -CH$_2$COOH 20:80	H	$\overset{}{\underset{O}{C}}CH_3$	80% 6-0-CM-CHITIN
$\overset{}{\underset{O}{C}}CH_3$	H	$\overset{}{\underset{O}{C}}CH_3$	MONOACETYL-CHITIN
$\overset{}{\underset{O}{C}}CH_3$	$\overset{}{\underset{O}{C}}CH_3$	$\overset{}{\underset{O}{C}}CH_3$	DIACETYL-CHITIN
CH$_2$CH$_2$OH	H	$\overset{}{\underset{O}{C}}CH_3$	GLYCOL-CHITIN (HE-CHITIN)
CH$_2$CHCH$_2$OH OH	H	$\overset{}{\underset{O}{C}}CH_3$	DIHYDROXYPROPYL-CHITIN (DHP-CHITIN)
PO$_3$H$_2$	H	$\overset{}{\underset{O}{C}}CH_3$	PHOSPHORYL-CHITIN (P-CHITIN)
SO$_3$H	H	$\overset{}{\underset{O}{C}}CH_3$	SULFONYL-CHITIN (S-CHITIN)
H	H	$\overset{}{\underset{O}{C}}CH_3$: H 70:30	70% DEACETYL-CHITIN (DAC 70)
CH$_2$CHCH$_2$OH OH	H	$\overset{}{\underset{O}{C}}CH_3$: H	DHP-CHITOSAN

Scheme 2. Chemical structures of chitin derivatives.

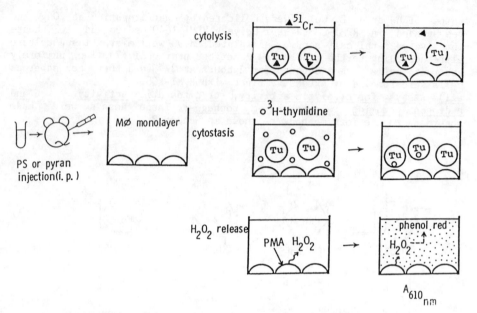

Scheme 3. Assay systems for macrophage activities.

Preparation of methamphetamine-6-0-CM-chitin conjugate(8)
D-N-(2-aminoethyl)methamphetamine(MAEA) was prepared by the condensation of D-methamphetamine with N-(2-bromoethyl)phthalimide in the absolute ethanol on refluxing for 12 hours. The product was refluxed again in the presence of hydrazine to remove phthalyl group as reported previously.
Resulted MAEA was condensed with 6-0-CM-chitin in deionized water by the use of Morpho-CDI at room temperature. About 30 moles of MA was estimated to be coupled per mole of 6-0-CM-chitin(M.W. 1.2×10^5 estimated from viscosity). 5mL of blood was collected from rabbit at various time intervals after the hypodermic injection of MA-CM-chitin saline solution and serum was prepared according to the previous method. The MA concentration in blood was estimated by ELISA method(Enzyme linked immuno sorbent assay) using working curve for the detection of MA in blood.

Inhibition of thrombin activity by chitin heparinoid(5)
Chitin heparinoids were prepared by the sulfation of various chitin derivatives under the condition reported previously.
Bovine antithrombin-III(AT-III)-heparinoid conjugate was applied to inhibit thrombin activity and also investigated on the inhibitions of another bovine coagulant factors.

RESULTS AND DISCUSSION

Activation of mouse peritoneal macrophages
The activations of mouse peritoneal macrophages were observed for partially deacetylated chitins and 6-0-CM-chitin as shown in Table 1.

There was little enhancement on the activation of macrophages with increase of pKa of amino group by the conversion to quaternary amine or by the connection of several amino acids. Anionic groups seem to be insensitive to the macrophage activation except carboxymethyl group. Though there was little difference on the macrophage activation between DAC-70 and 6-O-CM-chitin inducing by the dose of 500 ug derivatives at 3 days before harvest, 6-O-CM-chitin was more sensitive to the dose timing than that for DACs. This might be due to the higher biodegradability of 6-O-CM-chitin than that of DACs. However, a remarkable difference on the activation mechanisms between 6-O-CM-chitin and DAC 70 were shown by the production of cytokines. 6-O-CM-chitin induced only the activation of mitogenic activity through shortacting macrophage activation, whereas DAC 70 induced theproduction of cytokines except mitogenic activity for a fairly long period. A series of DACs, especially DAC 70, showed the stimulation of non-specific host resistance against Sendai virus and E. coli infections in mice. But there was few chitin derivatives to show this characteristic activity including 6-O-CM-chitins.

TABLE 1.
Activation of mouse peritoneal macrophages by various chitin derivatives.

Derivatives	Timing (days)	Dose (ug)	% Cytolysis (mean±S.E.)	% Stasis (mean±S.E.)
Chitin	-3	500	2.5 ± 3.0	58.2 ± 1.3*
6-O-CM-chitin(80%)	-3	500	53.0 ± 4.0**	58.3 ± 2.0*
P-chitin	-3	500	3.4 ± 4.2	55.6 ± 2.9**
S-chitin	-3	500	2.8 ± 6.6	43.1 ± 4.1
Acetyl-chitin	-3	500	0.3 ± 2.6	N. D.
6-O-HE-chitin	-3	500	22.2 ± 4.8*	44.1 ± 2.8
6-O-DHP-chitin	-3	500	15.3 ± 3.2	66.9 ± 1.5*
6-O-CM-chitosan	-3	500	3.9 ± 3.7	45.2 ± 3.7
6-O-DHP-chitosan	-3	500	18.6 ± 7.2	49.2 ± 2.1**
DAC 70	-3	500	55.9 ± 4.3**	54.7 ± 3.3**
Pyran copolymer	-3	500	66.4 ± 3.0**	97.5 ± 0.5*
Control			1.0 ± 5.6	38.9 ± 3.9

Chitin derivatives were injected intraperitoneally to mice at 3 days before(-3 days) harvest of periotneal macrophages.
(mean± S.E.): each value is mean ± standard error of six wells in each group.
* Statistically different from control by Student's t-test(p<0.05).
**Statistically different from control(p<0.005).

Lysozyme Susceptibility and Slow Release of Drug
The biodegradability of chitin has been observed to increase until C-3 hydroxyl group is substituted as shown in Figure 1. This enhancement of lysozyme susceptibility was suggested to be due to the destruction of crystalline structure on the specific substitution at C-6 hydroxyl group in addition to the increment of anionic binding sites and a remarkable decrease of susceptibility on the substitution of C-3 hydroxyl group was also suggested to loss of binding affinity for lysozyme(9). Then the biodegradable property was applied to drug delivery system, in which a sustained release of drug was intended by a hypodermic injection of drug

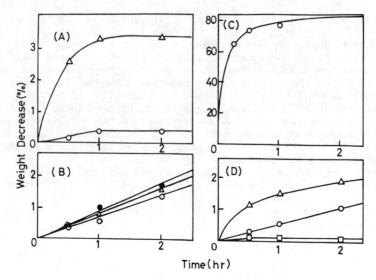

Figure 1. Time courses of lysozyme hydrolysis of chitin derivatives(insoluble fibers due to low degree of substitution). A: –△–; diacetyl-chitin fiber, –O–; mono-acetyl-chitin. B: –O–; n-butyl-, –△–; i-butyl-, –●–; t-butyl-chitin fibers. C: 6-0-CM-chitin fiber. D: –□–; chitin, –O–;dihydroxy-propyl-, –△–;glycol-chitin fibers.

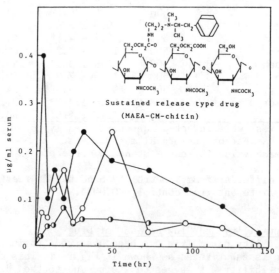

Figure 2. Slow release of "Prodrug" from drug bound 6-0-CM-chitin following to the hypodermic injection to rabbits. –●–; 3Kg of rabbit was treated with 4.2 mg of methamphetamine containing polymer conjugate. –O– and –◐–; 2Kg of rabbits by 2.8mg of drug containing polymer conjugate.

bound 6-O-CM-chitin saline solution. The blood level of drug bound CM-chitin oligomers(a prodrug) was maintained for fairly long period by a single hypodermic injection to rabbit as shown in Figure 2. The release of active drug would be achieved by the selective hydrolysis of "Prodrug"in the blood or organs, since the linkage between drug and spacer was not hydrolyzable one in this case(10).

Anticoagulant Activity of 6-O-CM-chitin Derivatives.

3,6-O- and N-sulfated 70% deacetylated 6-O-CM-chitin was shown to have the highest anticoagulant activity among sulfated CM-chitin derivatives with similar level as that of heparin as shown in Figure 3, when they were compared in the presence of antithrombin III(AT-III) and with the basis of sulfur content. Little hypertension of organ and body weights of mice was observed for 2 weeks following to the intravenous injection of heparinoid, whereas there were significant decrease of body weight and increase of organ weights by sulfated DAC-70 (2) as listed in Table 2. Only a mole of chitin heparinoid was shown to work most effectively through the formation of complex with AT-III among three moles of bound heparinoid.

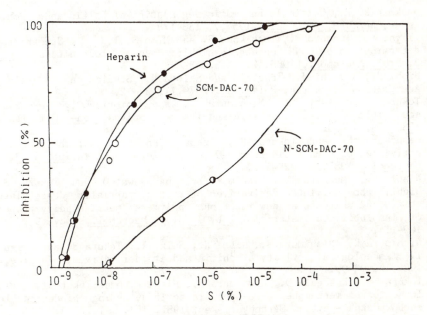

Figure 3. Inhibition profile of thrombin activity by 3,6-O- and N-sulfated, and N-sulfated 6-O-CM-chitosans through the formation of antithrombin-III-heparinoid complex.
SCM-DAC-70; 3,6-O- and N-sulfated 6-O-CM-DAC 70.
N-SCM-DAC-70; N-sulfated 6-O-CM-DAC 70.

The introduction of carboxymethyl group to DACs seems to be one of useful way to cancel their toxicity, because the activation of mouse peritoneal macrophages was also depressed by the introduction of carboxymethyl group to DACs(11).

TABLE 2.

Hypertension profiles by the intravenous injection of chitin heparinoids to mouse(2 weeks after the injection).

Treatment*	Dose (mg)	Ratio (died/tested)	Body weight(g)	Organ weight(%body weight)		
				Spleen	Lung	Liver
S-DAC-70	1	0/3	16.8	0.55	0.88	5.36
	2	0/3	16.9	0.71	0.91	4.97
SCM-DAC-70	1	0/3	17.3	0.35	0.73	4.68
	2	0/3	17.4	0.34	0.73	4.83
Control			17.4	0.34	0.75	4.83

*; Test solution(10mg/mL) was injected intravenously from tail vein of C57BL/6 mice.

REFERENCES

1. Hackman R. H., Chitin I. Enzymatic degradation of chitin and chitin ester, Aust. J. Bil. Sci., 1954,7, 168–178.
2. Miya, M., Iwamoto R., Yoshikawa S. and Mima S.,I.R. Spectroscopic determination of CONH content in highly deacetylated chitosan, Int. J. Biol. Macromol., 1980, 2, 323–324.
3. Nishi N., Noguchi J., Tokura S and Shiota H., Studies on Chitin I. Acetylation of chitin, Polym. J., 1979,11, 27–32.
4. Tokura S., Nishi N., Nishimura S. and Somorin O., Lysozyme-accessible fibers from chitin and its derivatives, Sen-i Gakkaishi, 1983, 39, 45–49.
5. Nishimura S. and Tokura S., Preparation and antithrombogenic activities of heparinoid from 6-0-(carboxymethyl)chitin,Int. J. Biol. Macromol., 1987, 9, 225–232.
6. Nishi N., Maekita Y., Nishimura S., Hasegawa O and Tokura S., Highly phosphorylated derivatives of chitin, partially deacetylated chitin and chitosan as new functional polymers: metal binding property of insolubilized materials. Int. J. Biol. Macromol., 1987, 9, 109–114.
7. Nishimura K., Nishimura S., Nishi N., Saiki I., Tokura S. and Azuma I., Immunological activity of chitin and its derivatives, Vaccine, 1984, 2, 93–99.
8. Tokura S., Hasegawa O., Nishimura S., Nishi N. and Takatori T., Induction of methamphetamine-specific antibody using biodegradable carboxymethyl-chitin. Carbohydr. Res.,1987, 161, 117–122.
9. Sashiwa H., Uraki Y., Saimoto H., Shigemasa Y. and Tokura S., Lysozyme susceptibility and substitution site by chemical modification of chitin. Proceedings 4th Int. Conference on chitin and chitosan, 1988, P-27, August, Trondheim, Norway.
10. Baba S., Uraki Y., Hasegawa O., Takatori T. and Tokura S., Controlled release of drug from carboxymethyl-chitin-drug conjugate.Proceedings XIth Int. Carbohydr. Symposium, 1988, p-388, August, Stockholm, Sweden.
11. Nishimura K., Nishimura S., Nishi N., Tokura S. and Azuma I., Effect of chitin heparinoids on the activation of peritoneal macrophages and production of cytokines in mice,in submission.

DISCUSSION

chaired by W. Burchard (Univ. Freiburg)

Q: A. Blažej (Slovak Tech. Univ.)
You have mentioned that chitin is biodegradable and chitosan is not biodegradable. I cannot agree with you. Chitin after deacetylation has free NH_2 groups and in acid medium is protonized and has positive charge, has character of an ion exchanger, has quite other conformation and is not cleaved by lysozyme, if you use glycosidase then is chitosan biodegradable. You cannot speak on non-degradable chitosan and degradable chitin. It is question of type of enzyme which you use.

A: You are right ! generally. But on the peritoneal macrophage activation study, deacetylated chitins which were more than 80% deacetylated, were almost undigestable by the glycosidases in macrophages. The chitosanases from bacteria are surely able to digest the chitosan. On my research, I am using "egg white lysozyme" as a model in most cases.

Q: R. Narayan (Purdue Univ.)
Is N-acetylated and 3-OH a sufficient condition for biodegradability even with completely substituted C-6 position ?

A: Theoretically yes. But it is impossible to prepare 100% substituted derivatives. Because there is not so much big difference on the reactivities between primary and secondary hydroxyl groups.

Q: G. Franz (Univ. Regensburg)
(1) Structure dependence (1-4 linkage compared to 1-3 linkages) with 70% deacetylated chitin.
(2) MW-dependence upon activity on tumor tested.
(3) Dose dependence on different tumors.

A: (1) Adjuvant activity of 70% DA chitin was almost similar level to that of lentinan.
(2) When M.W. of 70% DA chitin is reduced to 1/10~1/20, adjuvant activity did not reach original level as far as saccharide concentration is maintained as before.
(3) Yes, adjuvant activity depends on the sort of tumor cells.

ROUND TABLE I-1 \<Hi-Tech\>

"CELLULOSIC COMPOSITES"

Chairman: R. St. J. Manley
           (McGill Univ.)

INTRODUCTORY REMARKS

given by R. St. J. Manley

I feel greatly honored in having been invited to act as chairman of this session on Hi-Tech "Cellulosic Composites", and, like the previous session chairmen, I would like to take this opportunity to extend to Nisshinbo Industries my best wishes on the occasion of their 80th birthday, and also to congratulate Prof. Inagaki and the organizing committee for their great success in producing this splendid conference.

Polymer composites are now the forefront of advanced materials and an indication of their importance is shown by the fact that the American Chemical Society in a recent publication entitled "Opportunities in Chemistry" singled out polymer composites as a vital field for research and development.

In the area of cellulosics we can identify three types of composites, namely cellulose/matrix composites, wood/plastic combinations, and molecular composites (e.g. cellulose/synthetic polymer blends). In our symposium this morning we will have presentations on aspects of these types of cellulosic composites.

In the Round Table Discussion that will follow the lectures we will try to focus on the utilization of these materials in the foreseeable future, emphasizing potential uses and advantages and disadvantages in comparison with these types of composites.

CELLULOSE-BASED COMPOSITE FIBERS

Werner Berger, Mathias Keck and Bernd Morgenstern

Dresden University of Technology, Department of Chemistry,
Mommsenstr. 13, Dresden, GDR-8027

1. Introduction

On an international scale, increasing efforts are being made
to optimize the qualities of polymeric materials by dissolving
and mixing natural polymers such as cellulose, starch and
proteins with synthetic polymers such as polyacrylonitrile
(PAN) or polyamides in a joint solvent system. By spinning
these solutions, fibrous materials and films may be produced.
There are many patent information in this field. In contrast
to this, there are only few information published in this
field [1-4]. On the market, commercially produced fibres or
films and membranes are not yet available.

By this paper, it is intended to discuss from the scientific
point of view a basic position of the focal points of development taking into consideration literature and individual
findings. Analysing literatur and patent studies it is shown
that in order to make the appropriate mixed solutions the
total range of the newly found non-aqueous solvent systems for
cellulose such as

    paraformaldehyde / dimethyl sulfoxid (DMSO) [5]
    cyclic amine oxides [6]
    nitrosyl chloride / dimethyl formamide (DMF) [7]
    chloral / DMF [8]

were used, but respective works were not continued in these
systems for various reasons. In the solvent systems N_2O_4/DMF
and LiCl/dimethyl acetamide (DMAc), preferably PAN was used as
2nd component and extensive work was done in this field. Some
respective activities are shown in fig. 1.

The polymer mixing ratio was varied in wide limits, starting
at a low modification of PAN fibres with cellulose up to the

System	Author	Country	References
N_2O_4/DMF	Nakao	Japan	[9]
N_2O_4/DMF	Kaputskij	USSR	[10, 11]
N_2O_4/DMF	Schweiger	Belgium	[12]
N_2O_4/DMF	Prud'homme	Canada	[13]
LiCl/DMAc	Manley	Canada	[3]
LiCl/DMAc	Herlinger	FRG	[14]

Fig. 1: Preferably used solvent systems to make hybrid
         solutions

inverse ratio. According to the mixing ratio selected, with laboratory samples of spun fibres such properties as increased thermostability, tensile strength, dyeability, alkali resistance and resistance to light were obtained for PAN modified cellulose fibres [10] and such properties as water adsorption and dyeability were achieved for cellulose modified PAN fibres. In case of a feasible textile application, products of increased wearing qualities and/or increased hygienic properties could be expected. Similiar property modifications were obtained by protein modified PAN fibres [15,16] too. In the works oriented mainly towards application technology, there are only a few mechanistic statements on the nature of interactions in such systems. Kaputskij [10] explains the changed mechanic properties by a partial compatibility, mainly in the mixing range Cell : PAN = 90 : 10 and Manley [3] postulates the possibility of the formation of specific interactions of the type (1) without having them identified until now.

$$\begin{array}{l} \text{Cell} - \text{O} - \text{H} \\ \qquad \vdots \qquad \vdots \\ \text{PAN} - \text{C} \equiv \text{N} \end{array} \tag{1}$$

Proceeding from individual works on the nature of interactions in non-aqueous, non-derivatizing alternative solvents for cellulose [17] and appropriate model systems [18-22], we deemed it efficient to carry out investigations on the producibility of such metastable hybrid solutions and on the characterization of the possibly forming interactions in regenerated fibres and films, as it was expected that due to the thermodynamic incompatibility of both polymers in the coagulation process there is the possibility to obtain defined separation structures following specific conditions.

In the following, we understand hybrid solutions as the combination of at least two fibre-forming polymers in a common solvent where a natural polymer must be included.

2. Investigations in the system cellulose/PAN/LiCl/DMAc

Preparing hybrid solutions the two polymers are separately dissolved in LiCl/DMAc and the corresponding quantities of these solutions are mixed. As the run of cellulose dissolution and also the properties (e.g. viscosity) of the cellulose solutions are clearly influenced by the water content in the system, of course, the substances used should be carefully dried, in addition, protective gas is used. In our investigations wich are presented here preferably Heweten-201® (hydrolized linter with DP = 175 and α-cellulose content = 98 %) as well as Avicell-PH-101® (microcrystalline cellulose with DP = 165). The preactivation was performed following the solvent vapour method described by Terbojevich et al. [24]. In order to exclude the comonomer´s influence, in the first experiments a PAN homopolymerizate with a relativ solution viscosity of L_v = 2.15 and with an ash content of 0.1 % was used. Wolpryla-65® was used as polyacrylonitrile for spinning tests.

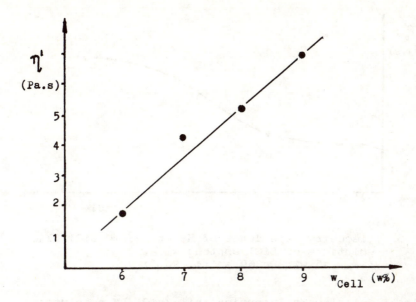

Fig. 2: Viscosity dependence of Heweten-201®/LiCl/DMAc
solutions on cellulose content w_{Cell}
(w_{LiCl}=7 w%; D=269 s^{-1}; ϑ=20°C)

Under the given preactivation and dissolution conditions [23],
solutions containing 6 - 7 w% of cellulose were reproducible
made. By the activation method applied, higher cellulose concentrations are obtained by experiments with great difficulties. The viscosity of the cellulose solutions increases with
increasing cellulose content (fig. 2). The viscosity is tremendously increased using a LiCl concentration exceeding 7 w%
(fig. 3), wich we explain by an increasing chain stiffening of
the cellulose by solvation.

For this reason to make hybrid solutions, the following conditions in the preparation of the cellulose solution were selected: LiCl concentration max. 7 w% and cellulose concentration
from 4 to 9 w%. The respective conditions for the PAN solution
induced a LiCl concentration of 7 w% and a PAN concentration
from 10 to 17.5 w%. At higher concentrations salting-out effects occur.

It turned out that under the given experimental conditions,
the homogeneity of the hybrid solutions (optical transparency
of the solutions as criterion for miscibility) depends on the
total polymer concentration and the polymer mixing ratio. At
a total polymer concentration less than 8 w% transparent solutions are obtained over the total mixing range. At a higher
polymer content this is the case when a Cell : PAN ratio of
2 : 98 or lower is adjusted. Hybrid solutions of a Cell : PAN
ratio of 5 : 95 or 10 : 90 respectively are unhomogenous where
phase separation is indicated by the opacifying of the solution. Cellulose solution used to make the F5 and F6 samples,

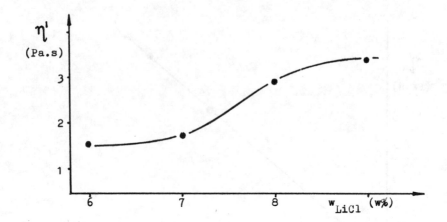

Fig. 3: Viscosity dependence of Heweten-201®/LiCl/DMAc
solutions on LiCl content w_{LiCl}
(w_{Cell}=6 w%; D=269 s^{-1}; ϑ=20°C)

Table 1: cellulose/polyacrylonitrile/LiCl/DMAc solutions

Sample	Solution composition Cell/PAN/LiCl/DMAc (w%)	Mixing ratio Cell : PAN (w / w)	Remark
F 1	0.06/ 9.90/7.0/83.0	1 : 99	clear
F 2	0.14/12.27/7.0/80.6	2 : 98	clear
F 3	0.14/14.70/7.0/78.3	2 : 98	clear
F 5	0.30/16.62/7.0/76.1	5 : 95	turbid
F 6	0.54/15.92/7.0/76.6	9 : 91	turbid
M 155	1.3 / 3.7 /6.0/89.0	25 : 75	clear
M 154	2.4 / 2.55/5.9/89.1	48 : 52	clear
M 152	3.7 / 1.25/5.9/89.2	75 : 25	clear
M 205	4.6 / 0.5 /6.0/88.9	90 : 10	clear
M 203	4.9 / 0.14/6.0/88.9	97 : 3	clear

show a non-Newtonian viscosity, i.e., in these cases cellulose
molecules are not completely solvated. Polymer-polymer interactions can occur resulting in a phase separation.

On the other hand we proved by means of viscosity measurements
[23] that for hybrid solutions which are homogenous the initial solutions show Newtonian flow behaviour. By this finding
the conclusion can be drawn that ideally solvated systems
where nearly all functional groups are saturated by specific
interactions with the solvent molecules form transparent hybrid solutions. This concept can only be applied for such systems where the polymer dissolution is bound to the formation
of specific interactions between the polymer and the solvent.
Phase separation of hybrid solutions occurs when the specific

polymer-solvent interactions are destroyed by temperature increase or by addition of precipitating agents in the coagulation process.

3. Interactions in cellulose/PAN solutions

Proceeding from these viscosimetric investigations in a hybrid solution no specific interactions between the polymers are expected to present, whereas interactions could be formed at the phase boundary layer caused by the phase separation arising in the coagulation process. We tried to obtain appropriate statements by spectroscopic investigations. Under this aspect experiments were carried out using solutions of the polymer combination cellulose/PAN in LiCl/DMSO. As the preparation of such solutions is not unproblematic methyl-β. D-glucopyranoside (MGluc) was used as model substance for cellulose. The results are listed in table 2.

Table 2: IR data of MGluc/PAN/LiCl/DMAc solutions
(n.n. not identifiable)

[MGluc]	[PAN]	[LiCl]	W_{PAN}	ν(OH)	δ(OH)	δ(OH)	ν(CN)	δ(CH2)
	(mol/l)		(w%)			(cm^{-1})		
MGluc in KBr			--	3380	1221	--	--	--
0.34	0	0	0	3325	1206	n.n.	--	--
0.25	0.99	0	52	3320	1209	n.n.	2243	n.n.
0.27	0	1.71	0	3290	1208	1316	--	--
0.72	0.96	1.74	28	3300	1209	1365	2242	1456
0.23	0.91	1.77	52	3300	1209	1362	2242	1456
0	0.91	1.77	100	--	--	--	2242	1455

After converting PAN and MGluc into the solvent most of the IR bands in their position change only slightly except the hydroxyl valence vibration of MGluc. For this vibration band a great bathochromic displacement of about 60 cm^{-1} is noticed. The reason for this is the solvation of hydroxyl groups with DMSO forming a hydrogen bridge of the type (2).

$$\text{Cell} - \text{O-H} \ldots \text{O} = \text{S} \underset{CH_3}{\overset{CH_3}{<}} \qquad (2)$$

This solvent effect is increased by introduction of LiCl into the solution. This can be identified by an additional displacement of ν(OH) band by 25 cm^{-1} (fig. 4)

The solvent's solvation influence on the nitrile vibration of PAN is considerably smaller. The band displacement is only 1 cm^{-1} in each case, but this is not surprising as the nitrile groups solvation occurs by the formation of minor EDA interactions. Within a test series with a changed MGluc/PAN ratio, the wave numbers of the ν(OH) and of the ν(CN) bands remain practically constant. The δ(CH2) vibration of PAN can be proved in the solution spectra but it is extremely wide and lies on

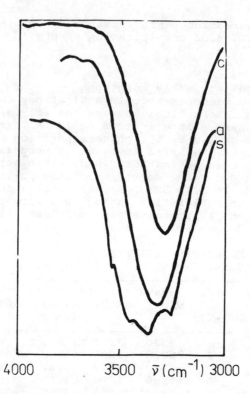

Fig. 4:   Hydroxyl valence vibration of MGluc
          (s... in KBr; a... in DMSO; c... in 1.8 M LiCl/DMSO)

on the hypsochromic flank of a DMSO absorption. Hence, a precise quantitative intensity determination is complicated. The
results obtained give no indication on specific polymer-polymer interactions in cellulose/PAN hybrid solutions.

4. Interactions in cellulose/PAN films

It seemed to be of importance for us to test whether this indication is changed by the coagulation of hybrid solutions.
For this purpose, casting films of Cell/PAN/LiCl/DMAc solutions of a thickness of 0.1 - 0.2 mm were regenerated in ethanol and the dried films were analyzed by IR-spectroscopy. The
film´s PAN content was varied from 3 - 75 w%. The wave numbers
determined for the characteristic vibration bands do not depend on the composition of the film within the error range.
For example, for the ν(OH) vibration 3415±5 cm^{-1} is obtained
and for the ν(CN) vibration 2244 cm^{-1}. In the mixtures spectra
no additional bands occur.

In order to varify whether the intensity of a vibration has
changed, the relative intensity I_{rel} of nitrile vibrations was
determined following the base-line method. Because the δ(CH$_2$)

$$I_{rel} = \frac{\text{extinction of } \nu(CN) \quad [2244 \text{ cm}^{-1}]}{\text{extinction of } \delta(CH2) \; [1455 \text{ cm}^{-1}]} \qquad (3)$$

vibration of PAN is not influenced by cellulose it was chosen as a standard. In fig. 5 I_{rel} is plotted against the polymer mixing ratio.

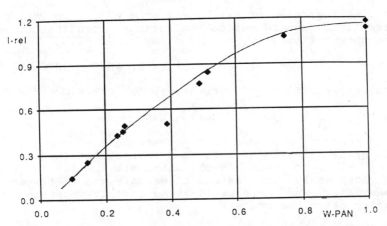

Fig. 5: Dependence of ther relative nitril band intensity I_{rel} on the PAN content W_{PAN} in Cell/PAN films

At a cellulose content of more than 25 % the intensity of nitrile bands considerably decreases. At a percentage of 15±5 % of polyacrylonitrile in the mixture the standardized intensity of the nitrile bands is less than half of the intensity of pure PAN. This means that in Cell/PAN blends at a PAN content from 10 - 20 % possibly about 50 % of the nitrile groups are included in specific interactions of the types (4).

$$PAN - CN \ldots H-O - Cell \qquad (4a)$$

$$
\begin{array}{c}
PAN - C \equiv N \\
\vdots \quad \vdots \\
Cell - O - H
\end{array}
\qquad (4b)
$$

This requires at the same time a definite phase strukture of the films. The domains of the phase rich in PAN should be comparatively small, so for PAN a great ratio of interface to phase volume is realized. Optically, this phenomenon is shown by the fact that Cell/PAN films of a PAN content up to 20 % have only a slight opacity. In contrast to this Cell/polyamide-6.6 films of the same composition are more intensily opacified. This can be easily identified by turbimetry and light microscopy.

5. Results of spinning tests and summary

Using a laboratory spinning equipment and stock solutions of the following composition:

	Polymer	LiCl	DMAc	
Wolpryla-65®	17.5	6	76.5	w%
Heweten-201®	6.0	6	88.0	w%

with a mixing ratio of PAN : Cell from 99 : 1 to 95 : 5 w% and taking into consideration the following conditions, we tried to spin test fibers:

Titer: 0.34 tex   Spinning bath: H_2O ; 22°C

Spinning jet: 600/0.07   Streching bath: DMF/H_2O = 50/50; 100°C

Streching: 1 : 6   Scouring bath: H_2O

As reference fibres, PAN pure polymerizate fibres (PAN-R) were used span of 17.5 w% solutions in LiCl/DMAc and fibres of a Wolpryla-I assortment (W-65) also. All the initial solutions were completely transparent after only one filtration stage. Spinnability and strechability definitely increased with increasing cellulose content.

In the non-stretched, gel-humid fibres a clear increase of radial capillars due to the coagulation is identified in the light microscopic cross-sectional takings, however, this is no more identified in stretched and dried fibres. The cellulose modified fibres are distinguished by an increase in water retention volume which is proportional to the cellulose content, in addition, by similiar streching and shrinking values of Wolpryla-65® and by improved dyeability. The tensile strength is below the standard values, this is caused by the use of a comparatively low molecular-weight cellulose (table 3).

Table 3: Physical properties of Cell/PAN test fibres

Ratio PAN:Cell	WRV (%)	Tensile strength (mN/tex)	Stretch (%)	Shrinkage (%)	Dyeability (%)
99 : 1	4.6	200	28.2	1.5	10.75
98 : 2	5.6	148	32.6	0.6	11.64
97 : 3	4.8	175	25.5	3.7	11.26
95 : 5	7.2	168	31.5	3.8	11.96
PAN-R	3.2	378	28.4	11.5	12.44
W-65	4.5	320	30.0	3.0	8.0

Summarizing the following statements are made:

1. There are no specific polymer-polymer interactions in hybrid solutions. Using initial polymer solutions with a Newtonian flow behavior hybrid solutions of an optional mixing

ratio can be made.

2. By making fibres from hybrid solutions the qualities of
   commercially produced fibres are varied, hence, their application is extended for other fields and uses.

3. Under the same conditions of coagulation, in different hybrid systems coagulation structures of different type,
   intensity and quantity are formed.

In future work we will try to investigate to what an extent a
controlled structure formation can be attained by defined
modification of polymer systems and conditions of coagulation.
It was revealed by first investigations (fig. 6) that a variation of the film morphology is obtainable in dependence on the
polymer mixing ratio. However, the possibility to make "composite membranes" could be promoted by a controlled structure
formation.

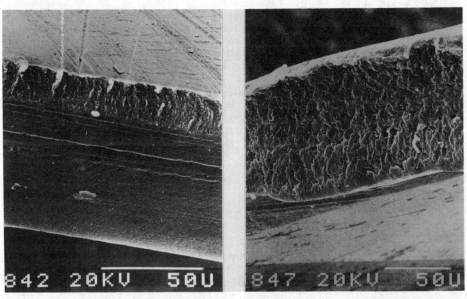

Fig. 6: Scanning electron micrographs taking of Cell/PAN films
(left with 25 w% PAN; right with 40 w% PAN)

6. References

[1] A.S. Tschegolja, F.N. Kaputskij, D.D. Grinshpan, E.Z.
    Burd and V.B. Kvasha; Lenzinger Ber. 51, 5 (1981)
[2] D.D. Grinshpan, F.N. Kaputskij, T.A. Savitskaya and V.A.
    Lyutsko; Acta Polymerica 36, 418 (1985)
[3] Y. Nishio, S.K. Roy and R.St.J. Manley; Polymer 28, 1385
    (1987)
[4] A.M. Jolan and R.E. Prud´homme; J.Appl.Polym.Sci. 22,
    2533 (1978)
[5] EP 19 566 (12.05.1980)

[6]   GB 1 144 048 (05.03.1969)
[7]   JP 46/7469 (24.02.1971)
[8]   JP 46/7468 (24.02.1971)
[9]   JP 45/2113 (24.01.1970)
[10] SU 1 002 419 (30.03.1981)
[11] D.D. Grinshpan, F.N. Kaputskij and T.A. Savitskaya;
      Zh.Prikl.Khim.(Leningrad) 50, 702 (1977)
[12] R.G. Schweiger; Tappi 57, 86 (1974)
[13] S. Savard, D. Lévesque and R.E. Prud'homme; J.Appl.
      Polym.Sci. 23, 1943 (1979)
[14] E. Pelz; Doctoral Thesis, Universität Stuttgart, 1987
[15] JP-OS 86 102 483 (20.10.1984)
[16] JP-OS 83 200 767 (18.05.1982)
[17] W. Berger, M. Keck, B. Philipp and H. Schleicher;
      Lenzinger Ber. 59, 88 (1985)
[18] W. Berger, M. Keck and B. Philipp; Cell.Chem.Techn.
      (in print)
[19] W. Berger, M. Keck, D.V. Shang and B. Philipp; Z.phys.
      Chem.(Leipzig) 266, 436 (1985)
[20] W. Berger, M. Keck, D.V. Shang and B. Philipp; Z.phys.
      Chem.(Leipzig) 266, 421 (1985)
[21] W. Berger, M. Keck, D.V. Shang and B. Philipp; Z.phys.
      Chem.(Leipzig) 267, 461 (1986)
[22] V. Kabrelian, W. Berger, M. Keck and B. Philipp; Acta
      Polymerica (in print)
[23] E. Fellmann; Diploma Thesis, TU Dresden, 1988
[24] M. Terbojevich, A. Cosani, G. Conio, A. Ciferri and
      E. Bianchi; Macromolecules 18, 640 (1985)

DISCUSSION

Comment: N. Shiraishi (Kyoto Univ.)
    Combination of PAN and cellulose will result in good composites.

A: Yes, it depends on the variation of the composition of the mixtures (solvents) and of the coagulation conditions.

Q: H. Inagaki (Kyoto Univ.)
    What kind of specific property do you expect for your composite fiber?

A: The variation of the composition of the mixture and of the coagulation conditions allows the production with special properties. One idea is to improve the H_2O absorption.

Q: A. Peguy (C. R. Macromol. Végét. CNRS)
    In your experiments have you studied the influence of the
coagulation conditions ?

A: Now we will investigate this as a very important problem
in the next time.

Q: K. Kamide (Asahi Chem. Ind. Co., Ltd.)
    (1) (Comment) We have established a theory of phase
separation of two chemically different multicomponent polymers dissolved in solvent, which gives solution critical
point, spinodal, cloud point curve (Polymer J., Nov.-issue)
    (2) Have you determined the solution critical point for
the solution ? And if yes, how did you do it ? You have
better to determine the two phase volume method.

A: (1) Excellent.
    (2) To determine the solution critical point for such
type of solutions is very difficult from the experimental
point of view. We investigate now this problem by light
scattering and another method. One point is very important.
- You need always inert conditions.

Q: T. Komoto (Gunma Univ.)
    I would like to recommend to use the ultra thin section
method to analyse the phase separation of the polymer blend,
using a transmission electron microscope. This is relevant to
Dr. Kamide's question.

A: Yes, I agree with your recommendation.

Q: W. G. Glasser (Virginia Tech.)
    What happens to crystallinity in the fiber ? Is it re-
tained ? Does it decrease ?

A: Not investigated now. I think it will be decreased. In
the future we will investigate this problem.

Q: D. G. Gray (McGill Univ.)
    You showed an infrared spectrum for cellulose in LiCl/
DMSO. Can you give any information on the relative effectiveness of this solvent compared to LiCl/DMAc ?

A: In the present time I don't have exact information about
your question. I will send you later this information.

Q: K. Tashiro (Osaka Univ.)
    Why the wavenumber of $C \equiv N$ streching mode is not affected
so much in the composite of cellulose and PAN system ? In
general, the position $\nu(C \equiv N)$ is very much affected by the
change of the environmental condition of the CN group.

A: In my opinion in the solution state are all -CN groups
solvated with the solvent. You now found a correlation. The
solvation influence on the nitrile vibration of PAN is considerably small. The band displacement is only 1 cm^{-1} in

each case. In PAN-Cell films there exists a dependence of the relative nitrile band intensity

$$I_{rel} = \frac{\text{extinction of (CN)} \quad /2244 \text{ cm}^{-1} /}{\text{extinction of (CH}_2) \ /1455 \text{ cm}^{-1} /}$$

on the mixing ratio.

NEW WOOD-BASED COMPOSITES WITH THERMOPLASTICS

CARL KLASON, JOSEF KUBÁT, and PAUL GATENHOLM
Chalmers University of Technology
Department of Polymeric Materials
S-412 96 Gothenburg, Sweden

ABSTRACT

Thermoplastics reinforced with cellulose fibres have been studied with
respect to the dependence of the mechanical properties on fibre concentration and fibre treatment. Prehydrolyzed fibres showing a high degree
of brittleness, were finely communited in the shear field of the processing machinery. The improvement in the mechanical stiffness and
strength was interpreted in terms of disintegration of the fibres into
submicroscopic microfibrils with high aspect ratios and high modulus and
strength values. The lack of dispersion of the fibres in the matrix normally found for untreated cellulose fibres was changed dramatically for
the prehydrolyzed fibres. They were finely dispersed as very short fibre
fragments and fibrils in the matrix after mixing in a twin screw compounding extruder. Also, good dispersion was obtained for fibres pretreated with a rubbery latex provided that the latex had good adhesion
to the matrix. Such treatment reduced the fibre attrition during mixing.
The findings of this investigation support the expectation that pretreated
cellulose fibres act as reinforcing fillers in thermoplastics.

INTRODUCTION

Cellulose and lignocellulosic materials show interesting potential as fillers
for thermoplastics. Although their use in thermosets is a well established
technique, the area of thermoplastics-based composites seems to have
attracted relatively little attention by processors and compounders.
However, during recent years there appears to be growing interest in
this area, at least judging from the increasing amount of relevant literature. Among the interesting properties of such fillers and reinforcing
fibres are their relatively low density and the possibility of pretreating

them in different ways (1-3).

This paper presents the results of an experimental study of the mechanical properties of composites consisting of prehydrolyzed cellulose and thermoplastic matrices. The main feature of prehydrolyzed cellulose fibres is their brittleness, permitting the fibres to be finely comminuted in the shear field of normal processing machinery. Such an effect is expected to improve the homogeneity and the mechanical parameters of the moulded samples.

The main part of the tests were done on injection moulded PP-samples containing varying amounts of prehydrolyzed cellulose fibres. Theoretical assessment of the moduli values using the Halpin-Tsai equation gave somewhat lower values than those recorded on experimental samples. This was interpreted in terms of the disintegration of the cellulose component into submicroscopic fibrillar entities, so called microfibrils, which can show stiffness and strength characteristics similar to those found for carbon fibres (1, 4). Also, the strength of the composites was found to increase with increasing fibre content when an appropriate coupling agent was added.

In another series of experiments suspensions of microfibrillar cellulose in water were prepared by intense mechanical treatment of prehydrolyzed cellulose. This fibrillar material was then incorporated into a thermoplastic resin by mixing it with a PVAC-latex. Films were prepared by drying thin layers of such mixtures. An optimum stiffening effect was recorded for pulps having DP-values between 200 and 400. The modulus value for a composite with 40 % microfibrillar cellulose was approximately four times higher than for a film containing untreated cellulose fibres. These experiments also show the inherent stiffening effect of cellulose microfibrils. When untreated cellulose fibres are used as reinforcement in thermoplastics significant fibre length reduction takes place in the compounding machine. Treatment of the fibres with an adhesion promoting system increases the dispersion of the fibres in the matrix and also diminishes fibre damage significantly. This is in good agreement with similar results found for short cellulose fibre reinforcement of rubber. In this way the inherent high aspect ratio of the ribbon shaped cellulose fibres can be fully utilized. The data obtained support the concept that cellulose has a significant hitherto largely unexploited potential as a reinforcing filler in thermoplastics.

Relevant literature data have been reported in connection with earlier publications on this subject (1, 5, 6).

EXPERIMENTAL

The polymer used was an injection moulding grade of PP (copolymer type, GYM 621 from ICI, $M_w = 83600$, as determined by GPC). Also, a PVAC-latex (anionic, Bg 389 from Borregaard) was used to facilitate the incorporation of the fibrils from a homogenized suspension (gel) into a polymer matrix.

A dissolving grade of cellulose (Ultra from Billerud, R18 94.2 %, R10 89.8 %, ethanol extract 0.2 %, ash 0.04 %, DP-value 764) was used. The hydrolysis was carried out using a 3 % oxalic acid solution in water at 0.2 MPa pressure and 121°C for 1 hour. The resulting DP-value was 239. After hydrolysis, the pulp was washed to pH 7, and dried at 70°C until a moisture content of 5 % was reached (3 days in an oven with recirculating air). The treated sample was disintegrated in a rotating knife mill (Rapid GK20, screen 4 mm) in order to facilitate feeding the compounding machine. After milling, the sample was dried at 100°C in vacuum in order to reduce the humidity to below 0.1 % prior to compounding.

The components were homogenized in a compounding extruder (Werner & Pfleiderer ZSK 30 twin screw extruder with a large number of kneading blocks, melt temperature 180°C). The extrudate strands were granulated before injection moulding (Arburg 221E/17R). The test bars (DIN 53455) with a cross-section of 10 x 3.5 mm and an effective length of 75 mm were conditioned at RT and 50 % relative humidity for 24 hours before testing at RT (deformation rate $4.5 \cdot 10^{-3}$ s^{-1}). From the stress-strain curves, the tangent modulus E, the stress at yield and rupture and the corresponding deformations were determined. The impact strength was measured with a Frank instrument (model 565 K, Charpy, DIN 53453).

In a series of experiments the prehydrolyzed fibres were premixed with $CaCO_3$ particles (average size 1 μm, properties 1:1) and a coupling agent (maleic anhydride modified PP, 10 % of the filler amount) in the extruder (die removed, cylinder temperature 100°C) in order to make an intimate mixture of the treated fibres, the minerals particles and the coupling agent prior to compounding.

Commercial cellulose fibres treated with a butadiene latex (Santoweb W from Monsanto) were compounded with a blend of PP with 10 % SEBS block copolymer (Shell, type G 1652, nominal MW_m 7-40-7) in the twin screw extruder. These experiments were performed in order to evaluate the importance of the fibre to matrix interaction on the dispersion of the fibres.

RESULTS and DISCUSSION

Figure 1 shows the relative modulus, defined as the ratio of the composite modulus to the matrix modulus, at different fibre contents for PP filled with untreated (unfilled symbols) and prehydrolyzed cellulose fibres (filled symbols). The data for the hydrolyzed fibres mixed with an equal amount of $CaCO_3$ is also shown in figure 1. The untreated fibres give a very low reinforcing effect, mainly due to the fibre length reduction and uneven dispersion of the fibres. The positive influence of the hydrolytic treatment on the increase in stiffness is clearly seen. For the case, when the treated fibres were premixed with $CaCO_3$ and a coupling agent prior to compounding, a pronounced increase in modulus was found.

Improved distribution of the coupling agent due to the premixing step is a probable explanation of the increase found with $CaCO_3$-blending. Also, the mineral particles may act as grinding elements. When the coupling agent is added directly with the PP resin in the extruder, no improvement in stiff-

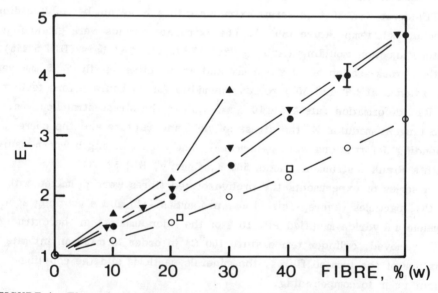

FIGURE 1. The relative modulus vs. cellulose fibre concentration in PP. Fibre symbols: (o) untreated, (●) prehydrolyzed, (▼) prehydrolyzed with coupling agent, (▲) prehydrolyzed with coupling agent and premixed with $CaCO_3$, and (□) Santoweb.

ness is noted in comparison to the prehydrolyzed fibres without coupling agent. However, the dispersion of the fibres is very good.

The Santoweb fibres in PP/SEBS-blend give a stiffening effect of the same order as the untreated fibres, figure 1. There is a fundamental difference between these two samples; the dispersion of the fibres is very low for the untreated fibres, whereas the Santoweb fibres are evenly distributed throughout the matrix. This is due to the presence of the SEBS phase, which acts as an adhesion promoting agent between the butadiene coated fibres and the PP matrix. Without the SEBS phase, the Santoweb fibres shows poor dispersion in PP.

The average fibre dimensions in the moulded test bar were determined for the samples shown in figure 1, and expressed as aspect ratios. They were approximately 4 and 7 for hydrolyzed and untreated fibres, respectively. The aspect ratio of the Santoweb fibre was above 50. It has to be emphasized that the microfibrils cannot be seen in the microscope used. Thus, the composites with the hydrolyzed cellulose may contain a certain fraction of these reinforcing microfibrils.

The modulus is one of the properties of a composite containing a fibrous or particulate phase which can be calculated in a relatively easy way using the theoretical model of Halpin-Tsai or its simplified form proposed by Lewis-Nielsen (7):

$$E_c/E_1 = (1 + ABv_2)/(1 - Bpv_2) \qquad (1)$$

where E_c and E_1 denote the elastic moduli for the composite and the matrix, respectively, and v_2 is the volume fraction of the filler. The factor p takes into account the packing fraction of the filler, v_m, at the maximum filler concentration and is expressed as $p = 1 + (v_2 - v_2 v_m)/v_m^2$. The value of A, which is related to the Einstein coefficient k_E through $A = k_E - 1$, has the value of 1.5 for rigid spheres and could be expressed as $A = 2(l/d)$ for fibres with aspect ratio l/d. For the composites containing both fibres and $CaCO_3$, the Halpin-Tsai equation is applied to each type of filler, and the E_c values are then weighed together using the method described in an earlier paper (6). The estimated E_c values are given in table 1 for a cellulose fibre content of 30 %. A comparison with experimental data is also made.

TABLE 1

Prediction of E_c/E_1 of the composites according to equation (1) and comparison with experimental data. Fibre concentration 30 % by weight. E_1 = 1.5 GPa.

	(E_c/E_1)theory	(E_c/E_1)exp	Aspect ratio
Untreated	2.6	2.0	7
Hydrolyzed	2.3	2.6	4
Hydrolyzed/$CaCO_3$ (1:1)	3.0	3.8	4
Santoweb	3.6	2.6	50

Table 1 contains some irregularities. The untreated fibres and the Santoweb fibres produce lower moduli than expected in view of their aspect ratios. For the hydrolyzed variety, the opposite is true. Further, the fibres had a low degree of orientation. When this is taken into account the theoretical E-values show fair agreement with the experimental data for the Santoweb fibres and the untreated fibres. The high stiffening effect of the hydrolyzed fibres may be due to a partial comminution of the fibres into microfibrils.

The reinforcing potential of microfibrils was demonstrated in the following way. A dispersion of microfibrils in water was prepared by exposing pre-hydrolyzed cellulose to the intense shear force of a slit homogenizer. This gel was mixed with an emulsion of PVAC particles in water, and made into a film. The E-modulus was found to increase from 63 MPa for unfilled PVAC film to 2400 MPa at a fibre content of 40 %. The corresponding modulus for a film containing untreated cellulose was 650 MPa.

The results showing the variation of the yield stress value with the filling level are shown in figure 2. On the whole, the yield stress value decreases with increasing cellulose concentrations, except for the samples containing a coupling agent (maleic anhydride PP-copolymer), where an increase was noted. The most pronounced reinforcing effect was found for the samples filled with the precompounded mixture of $CaCO_3$ and fibres. The reason for this is probably that the coupling agent, which was present during the premixing, was well dispersed onto the fibres.

The stress at rupture depends on the fibre concentration in a way similar to the yield stress values.

The elongation at rupture and the impact strength values showed a marked decrease with increasing fibre concentrations. The impact strength dependence on fibre concentration was essentially the same for hydrolyzed and untreated fibres. This is evident from figure 2 (right). There was a marked increase in impact strength for the Santoweb fibres. This is a result of the addition of an elastomeric component to the matrix (SEBS, 10 %), and probably also due to the flexible butadiene layer on the Santo-web fibres. Results similar to those presented above are found when other thermoplastics like HDPE, LDPE and PS are used as matrix materials.

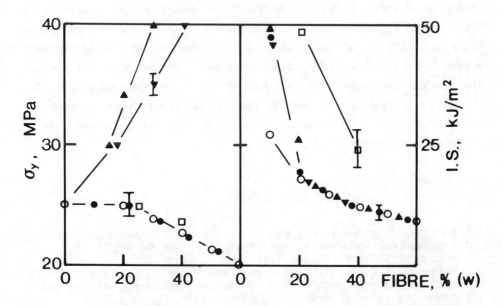

FIGURE 2. Stress at yield (left) and impact strength (right) vs. fibre concentration for PP composites. Fibre symbols: same notation as in fig. 1.

CONCLUSIONS

The main result of this report is the demonstration that a hydrolytic treatment of the cellulose fibre and the resulting embrittlement produces significant improvement in the modulus of the corresponding PP-based composites. The strength and to some extent also the stiffness is greatly improved when an appropriate coupling agent, giving good adhesion between the fibres and the matrix, is included. The E-values recorded for such samples exceed the values obtained with carefully compounded, unhydrolyzed fibres. Although the majority of the embrittled material in the matrix appears to occur in the form of irregular fragments, a disintegration into high-modulus, high-strength fibrils is discernible. The poor dispersion found for untreated fibres is significantly improved for the prehydrolyzed fibres. Good dispersion can also be achieved, if the fibres and the matrix show good adhesion to each other. This was demonstrated for butadiene coated fibres (Santoweb) in a SEBS-modified PP matrix. With this system, the inherent high aspect ratio of the cellulose fibres could be preserved. The lack of stiffening observed for these fibres is mainly due to their isotropic orientation in the moulded test bars. This can be improved by chosing a suitable processing technique.

REFERENCES

1. Boldizar, A., Klason, C., Kubát, J., Näslund, P. and Saha, P., Prehydrolyzed cellulose as reinforcing filler for thermoplastics, Intern. J. Polymeric Mater., 1987, 11, pp. 229-262.
2. Maldas, D., Kokta, B.V., Raj, R.G. and Daneault, C., Improvement of the mechanical properties of sawdust wood fibre-polystyrene composites by chemical treatment, Polymer, 1988, 29, pp.1255-65.
3. Kishi, H., Yoshioka, M., Yamanoi, A. and Shiraishi, N., Composites of wood and polystyrenes, Mokuzai Gakkaishi, 1988, 34, pp. 133-139.
4. Battista, O.A., Microcrystal Polymer Science, McGraw-Hill, New York, 1975, pp. 1-57.
5. Klason, C., Kubát, J. and Strömvall, H.-E., The efficiency of cellulosic fillers in common thermoplastics, part I. Intern. J. Polymeric Mater., 1984, 10, pp. 159-184.
6. Dalväg, H., Klason, C. and Strömvall, H.-E., as above, Part II, Intern. J. Polymeric Mater., 1985, 11, pp. 9-38.
7. Nielsen, L.E., Mechanical Properties of Polymers and Composites, Marcel Dekker, New York, 1974, pp. 453-478.

DISCUSSION

Q: S. Kohjiya (Kyoto Inst. Tech.)

(1) Your presentation was mainly on the physical properties of the composites, but my question is on the process-
ability. Generally speaking, the more reinforcing the filler,
the poorer is the processability. How about on your composites ?

(2) Did you find some excellent one among your coupling
agents that improves both the physical properties and the
processability ?

A: (1) The improved melt strength and the improvement in
dispersion found for the pre-hydrolyzed CTMP-fibers make
processing more favorable compared to untreated fibers,
which make agglomerates. The increase in viscosity is moder-
ate and can be compensated for during processing.

(2) The opinion is that maleic anhydride(MA-PP), and also
some isocyanides, acts as very good coupling agents. Silane
coupling agents do not give good results; it is reported that
some titanate type can give some improvements in strength,
but until now we have not found a better choice than the MA-
PP type.

Q: W. Berger (Tech. Univ. Dresden)

In the second or third slide you discuss the correlation
between E-modulus and different fiber material. Did you find
any influence of the fiber length ?

A: From the Halpin-Tsai equation, the E-modulus is increasing with an increase in fiber length (or aspect ratio). We
have tried the following pulps (softwood): bleached sulphate,
sulphite, dissolving pulp and birch sulphite. The best results, with regard to the mechanical parameters, were found
for the bleached sulphate pulp (50% glue 50% spruce).

Q: J. J. Silvy (Ec. Franc. Papeterie)

Would you please clarify my mind on the effect of orientation of the fibers in the matrix during the phase of extru-
sion ? You said that there was sometimes a negative effect
in respect of the curl of the fibers and in other case that
the material is quite isotropic with particles of high shape
ratio. Isn't some orientation effect of the particles in this
process ?

A: For the straight and stiff unmodified wood or cellulose
fibers, there is an orientation in the direction of flow.
However, for the commercial fibers ('Santoweb'), covered with
an elastomeric layer, there was a general tendency that the
fibers showed a low degree of orientation. Of course, this
can be improved by a better control of the processing parameters.

Q: H. West (Univ. Col. North Wales)

What is the influence on the toughness of your composites

of improving the compatibility between the fiber and matrix ?

A: The toughness of a composite depends on the nature of the interface between the fiber and the matrix. If this interface is soft, a high impact resistance will result. However, a perfect coupling between filler and matrix may give rise to low impact resistance if the interface is too hard.

Comment: S. Kohjiya (Kyoto Inst. Tech.)
    About 10 years ago a chemical company in Japan placed a question. They were working wood/plastics composites to be used as automobile parts. But, it was found that the shape stability of the products were not good enough to meet the specifications of auto industries. My personal impression is that they may find applications as interior parts of automobiles.

A: Yes, I think one has to use these materials for interior parts of automobiles. I think that the material characteristics have to be changed somewhat: the dimension stability due to moisture sorption must be increased, and the impact resistance at low temperature must increase as well.

Q: A. Å. Johansson (Tech. Res. Cent. Finland)
    Thank you for your very interesting paper. Could you please comment on the influence of humidity (in spite of prior drying of fibers) at the interface between fiber and matrix, (on strength properties) ?

A: Moisture sorption tests show that moisture content is proportional to filler concentration. This will cause some problems with the dimension stability of large panels used by the automotive industry. However, moisture sorption can be reduced by adding maleic anhydride modified-PP, or by cross-linking of the wood fibers (acetylation). It is very important that all moisture of the filler is removed before or during processing, otherwise the adhesion between the fibers and the matrix will be poor. Also, a lot of voids are formed in the finished product.

PLASTICIZATION OF WOOD AND ITS APPLICATION

NOBUO SHIRAISHI

Kyoto University/Japan

ABSTRACT

Through chemical modifications, wood has been converted to a thermally flowable material. In this connection, it should be emphasized that thermoplastic properties can be conferred on wood by the internal plasticization through the chemical modifications. In some cases, however, thermoplastic properties must be supplemented by external plasticizations through blending with plasticizers. Such plasticization of wood can provide us with ample opportunities in cultivating wood processing and utilization of wood. Relating to this, not only the chemically modified wood but also untreated wood were recently found to dissolve in various solvents and solutions. This liquefied wood solution and plasticized wood have been applied to study for the followings: preparations of films, sheets and other moldings, preparations of three-dimensionally cured plastic-like wooden boards, hot-melt adhesives and deep-drawable hardboards, surface-layer plasicization of wood, preparations of wood-based reactive adhesives, foams, fibers and carbon fibers. Obtained results in these lines of works were introduced and future potential of wood plasticization was discussed.

INTRODUCTION

Since the advent of our human beings, wood has been used for our life. This is because wood is a convenient material; it can be sawed, planned, and driven with a nail. In addition, it has several remarkable features as a building material.

Wood is composed of cellulose (50 - 55 %), hemicellulose (15 - 25 %) and lignin (20 - 30 %), with small quantities of ash and extractives. The main components make up an interwoven network in the cell walls and middle lamella. Within the cell walls, lignin is believed to exist as a stereoscopic sponge-like aggregate. Through the network of this structure of lignin, cellulose is interwoven as fibril bundles, and the spaces are filled with hemicelluloses. The crystalline portions of cellulose fibrils can be regarded as rigid junctions among the linear polymer chains. In this sense, cellulose fibrils must form a three-dimensional network structure as well. The cell walls can, therefore, be considered as an interpenetrating polymer network (IPN) with chemical interactions between the main components of wood, such as those in lignin-carbohydrate-complexes (LCC). Our recent evidence is suggesting that the LCC is acting as a compatibilizer in the cell wall, by being localized at the interface between cellulose and lignin macromolecules. This must improve the adhesion of these wood components each other[1]. All these facts imply that wood is an excellent natural composite .

CULTIVATION OF WOOD PROCESSING BY PLASTICIZATION

Wood is not included in the "three main materials" since the method of processing is very limited for wood. However, metals, plastics and glass can be processed under liquid phase at high temperatures. Such differences are, therefore, due to a lack of plasticity in wood, meaning that it can not be melted, dissolved or softened sufficiently for molding.

The methods in wood processing are so limited at present that a scope of its utilization is restricted. This often makes wood itself less valuable as a material. However, once plastic properties are added to wood, it becomes more useful material . This further broadens the method in wood processing to a variety of fields. In this way, wooden material which is limited in use can now be modified into a high quality product with additional value. Furthermore, utilization of wastes from wood, for example, would be made viable.

Such a new technique has been highly evaluated recently and it is now evident that wood is converted into materials with the same features as plastics. For example, through esterification with higher fatty acid or etherification such as benzylation, thermoplastic moldable properties were found to be conferred to wood (Fig. 1). Furthermore, dissolution or liquefaction of wood has become possible (Fig. 2). This means that wood will become more usable for the production of valuable materials such as plastics, in the future.

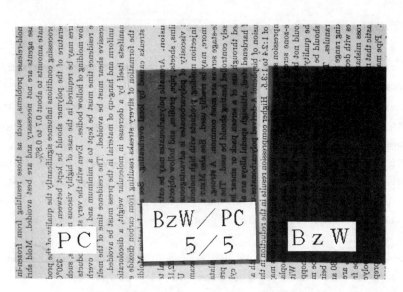

Figure 1.   Films prepared from benzylated wood meals (right), from an equal
weight blend of the benzylated wood meals and polycarbonate
(center), and from the polycarbonate (left).

Figure 2.   A phenol solution of carboxymethylated wood.

CELLULOSE CRYSTALLINE PREVENTING WOOD FROM THERMAL FLUIDITY

Let us start from a subject "why wood does not have plasticity". As mentioned above, cellulose molecules within wood form bundles. They also form a crystal structure, the proportion of which reaches about 70 %. This is because cellulose is a linear and streoregular polymer composed of glucose residues as a repeating unit.

It is known that the melting point of crystalline cellulose is considerably higher than its decomposition temperature. Thus, cellulose would be a material with low thermoplasticity due presumably to its crystalline structure. This is one of the main reasons that wood cannot reveal thermal fluidity.

However, once cellulose is derivatized by chemical modifications, the thermal properties of cellulose can be altered as seen in cellulose nitrate, cellulose acetate, benzyl cellulose and so forth as cellulose plastics. In this sense, it could be easily postulated that if cellulose in wood is derivatized in situ, wood itself would gain thermoplasticity.

Nevertheless, no one has thought to make such a trial until quite recently. This is rather natural because people have been believing the accepted concept for last forty years that a lignin molecule has a three-dimensional gel structure, preventing derivatized wood from being thermally flowable. In 1979, however, the idea of converting the whole wood into plastics has appeared[2].

MORE EFFECTIVE THERMOPLASTICIZATION OF WOOD BY A LARGER SUBSTITUENT

The author and coworkers have found recently that thermoplastic properties can be conferred to wood by chemical modification such as esterification, etherification and some other derivatization[2]. Figure 3 shows an example of thermally flowable wood. It is shown that lauroylated wood meal (a) could be molded to a film (b) by hot-pressing. Although the colour of the film is pale yellow, it is transparent. Benzylation of wood also gave the same result (Fig. 1).

These results are due to an introduction of a substituent group which enables wood to be internally plasticized. Therefore, derivatization by large substituents, such as the lauroyl $CH_3(CH_2)_{10}CO-$ and benzyl ($C_6H_5CH_2-$) would be one of the easiest ways of conferring thermoplasticity to wood.

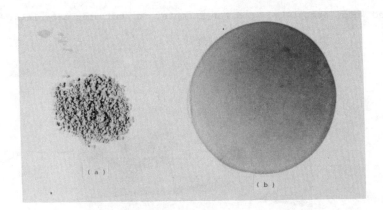

Figure 3. Lauroylated wood meals (a), and a film prepared from the lauroylated wood meals by compression molding.

There is a regularity in conferring thermoplasticity to wood. When a small substituent group and/or polar group is introduced, thermal fluidity could not be achieved. This means that such a derivatization can not affect wood to obtain plastic properties. The lack of plasticity was found to be supplemented by additional external plasticization. For example, allylated wood which did not show complete thermal fluidity, was altered to be a flowable and moldable material as blended with certain amounts of polymethylmethacrylate (PMMA) and dimethylphthalate. From these observations, it can be concluded, in principle, that irrespective of the molecular size of substituent groups, thermoplastic properties can be achieved by internal plasticization through chemical modification. If the effect is insufficient, it can be supplemented by external plasticization.

APPLICATION OF THERMOPLASTICIZATION OF WOOD

As described above, it has become possible to confer thermofluidity to wood. Based on these findings, several attempts have been made for its application. Examples of this are (1) preparation of films, sheets and other moldings from the chemically modified wood (Figs. 1 and 3), (2) preparation of three-dimensionally curred plastic-like wooden boards, (3) preparation of deep-drawable hard board (Fig. 4), (4) application to hot-melt adhesives, (5) surface-layer plasticization of wood intended to develop surface-densified and embossed materials, and (6) chemically modified wood, acetylated wood for example, for building materials.

Figure 4. Hot-press drawability of hard board with the density of
approximately 1.0. Ordinary hard board (left) and hard board
prepared from acetyl-propionylated wood pulp (right).

There have been many trials in preparation of films, sheets and other
moldings. Moldings with various physical properties could be obtained from
chemically modified woods and their blends with synthetic polymers. For
example, mechanical properties of the films from benzylated wood are
completely comparable with those of common synthetic polymers. The
mechanical properties of the benzylated wood film can be further enhanced
by blending with polystyrene (PS), in which styrene-maleic anhydride
copolymer (SMA) is added in a samll amount. Presence of small amounts of
the SMA also improves the tensile properties of blended films from
benzylated wood and modified polyphenylene oxide.

From a standpoint of preparing moldings with enhanced properties,
three dimensionally cured plastic-like wooden boards have been prepared from
chemically modified wood[3]. This is the case of thermo-set molding and the
important point is that of making use of chemicals which can act as
plasticizers for the modified wood, and at the same time can react with the
component of the modified wood during the hot-press moldings. For instance,
phthaloylated wood as an example of carboxy-group-bearing esterified wood
was used as chemically modified wood; whereas bisphenol A diglycidyl ether
as the plasticizer and the crosslinking agent. In the presence of the
latter, the thermally flowable molding can be made with cross-linking

reaction during the hot-press molding stage. The prepared board containing 60 % of wood has compressive strength as high as 2000 kgf/cm^2. Other properties such as hardness, thermo-moldability and water resistance are also all excellent.

An example of a three-dimensionally moldable fiber boards, or deep drawable fiber boards, is shown in Fig. 4. In this case, wood fibers were just chemically modified to a level to make the products thermoplastic but not thermally flowable. Fiber boards that are highly moldable are shown to be obtained.

DISSOLUTION OR LIQUEFACTIION OF CHEMICALLY MODIFIED WOOD

Chemically modified woods have been found to dissolve or liquefy in neutral aqueous solvents, organic solvents or solutions, depending on the characteristics of the modified wood. So far, three methods have been found for wood liquefaction.

The first trial of the liquefaction of wood was accomplished by using very severe dissolution conditions. One example is that series of aliphatic acid esterified wood samples could be dissolved in benzyl ether, styrene oxide, phenol, resorcinol, benzaldehyde, aqueous phenols, chloroform-dioxane mixture and so forth after dissolution treatment at 200-270 $^{\circ}$C for 20-150 min.

Another method of liquefaction is to make use of solvolysis during the process. By using conditions which allow phenolysis of a part of lignin, especially in the presence of an appropriate catalyst, the liquefaction of chemically modified wood into phenols could be accomplished under milder conditions (at 80°C for 30 to 150 min). Allylated wood, methylated wood, ethylated wood, hydroxyethylated wood, acetylated wood and others have been found to dissolve in polyhydric alcohols such as 1,6-hexanediol, 1,4-butan-diol, 1,2-ethanediol, 1,2,3-propane triol and bisphenol A, by use of dissolution conditions described above.

The dissolution processes can give rise to paste-like solutions with considerably high concentration of wooden solute (70 %). Figure 5 shows an earlier stage of the dissolution. Dissolution has just proceeded for 5 min after adding carboxymethylated wood meals into equal weight of phenol in the presence of catalyst, hydrochloric acid, maintained at 80 $^{\circ}$C. Figure 2 shows a solution of carboxymethylated wood in phenol with a concentration of 50 %. Almost complete dissolution could be confirmed.

Figure 5.   The early stage of dissolution of hydroxyethylated wood meals into phenol.

For the third method of dissolution, post-chlorination has been developed[4]. When chemically modified woods are chlorinated, their solubility to solvent is enhanced tremendously. For example, at room temperature, cyanoethylated wood can dissolve in o-cresol only 9.25 %. However, once chlorinated, it can dissolve almost completely in the same solvent.

APPLICATIONS OF DISSOLUTION OR LIQUEFACTION OF CHEMICALLY MODIFIED WOOD

There are many potential applications in dissolution or liquefaction of chemically modified wood. Examples of application include (1) fractionation of modified wood components, (2) preparation of solvent-sensitive and/or reaction-sensitive wood based adhesives, (3) preparation of resinified wood based moldings such as foam type molding, (4) preparation of wood based fibers and their conversion to carbon fibers.

For the fractionation of modified wood components, dissolution-precipitation technique has been successfully used.

The preparation of adhesives from chemically modified wood has been studied. In these cases, phenols, bisphenols and polyhydric alcohols have been used as solvents for the modified wood, and the resins should contain meaningful amounts of the modified wood.

Combined use of these reactive solvents with reactive agents, cross-linking agent and/or hardener, if necessary, have given rise to phenols-formaldehyde resins (such as resol resin), polyurethane resins, epoxy resins etc., all of which contain a meaningful amount of chemically modified wood and reveal excellent gluability.

Molding materials such as foams or shaped moldings can be also obtained from the chemically modified wood solutions. One of the examples is shown in Fig. 6. This can be prepared by adding an adequate amount of water as a foaming agent and a polyisocyanate compound as a hardener, to the 1,6-hexandiol solutions of allylated wood, mixed well and heated. When heated at 100°C, foaming and resinification of resins are initiated within 2 minutes and completed within several minutes. If promotors such as triethylamine are added, rapid reactions occur even at room temperature and foams can be obtained within several minutes. The foam shown in the figure has a low apparent density of 0.04 g/cm^3, a substantial strength and restoring force for the compression deformation.

Figure 6.   Appearance of polyurethane foams from allylated wood.

One other application of modified wood solutions is the formation of filament or fibers[5]. After getting the phenol solution of the acetylated wood, hexamethylene tetramine is added and heated up to 180 °C to promote addition-condensation for a resinified solution with high spinnability. From the solution, the filaments are spun and hardened in a heating oven at a definite heating rate. Maximum temperature for the hardening is 250 °C. By this way, endless filaments can be easily obtained. Figure 7 is a good example of the filament obtained by spinning.

These filaments can be carbonized to give carbon filaments. The strength of the carbon filaments was measured according to Japan Industrial

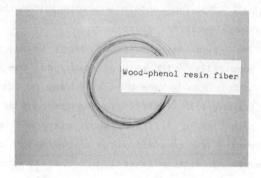

Figure 7. Appearance of the acetylated wood-phenol resinified filament.

Standard and the tensile strength up to 100 kgf/mm^2 has been obtained so far. This strength is comparable to that of the pitch carbon fibers of general purpose grade. Further improvements of its strength will be expected by improving the methods of spinning and carbonization.

DISSOLUTION OR LIQUEFACTION OF UNTREATED WOOD

So far, dissolution or liquefaction of chemically modified woods has been introduced. However, more recently, untreated wood was found to be dissolved or liquefied also in several organic solvents[6]. For example, after the treatment at 200 - 250 ^{o}C for 30 - 180 min, wood chips and wood meals were dissolved in phenols, bisphenols, alcohols (benzyl alcohol), polyhydric alcohols (1, 6-hexanediol, 1, 4-butanediol) and oxyethers (methyl cellosolve, ethyl cellosolve, diethylene glycol, triethylene glycol, polyethylene glycol), 1, 4-dioxane, cyclohexanone, diethyl ketone, ethyl n-propyl ketone.

Obtained liquefied wood is paste-like solutions with highly concentrated wooden solute as high as 70 %. After the dissolution, wood components were found to be certainly degraded and became reactive. Thus, obtained wooden solute can be used to prepare adhesives and other moldings. This will open a new and practical field for utilizing wood materials.

APPLICATION OF THE DISSOLUTION OR LIQUEFACTION OF UNTREATED WOOD

From wood solutions of untreated wood, almost the same products have prepared as those from the chemically modified wood. For example, resol-type phenol resin adhesives prepared from five parts of wood chips and two parts of phenol did not require severe adhesion conditions and were comparable to the corresponding commercial adhesives in their gluabilities. Acceptable waterproof adhesion was attained for the adhesives after gluing wood veneers at 130 $^\circ$C.

As another example of the application, foams can be prepared from untreated wood – polyethylene glycol solutions. Soft-type (Fig. 8) and hard-type foams can be prepared by changing the preparation conditions. The prepared foams were of density around 0.04 g/cm^3, having substantial strength and restoring forces for deformations. These facts imply that wood components are not merely blended within the foams but also play an important role extensively in holding shapes or keeping dimensional stability of the foams.

Figure 8. Appearance of polyurethane foams prepared from untreated wood after liquefied by polyethyleneglycol 1000.

The carbon fibers described previously could be also prepared from the untreated wood solutions. Tensile strengths of up to 120 kgf/mm^2 has been obtained so far. Further improvements of their physical properties will be also expected.

CLOSING

The present stage of studies on wood plasticization has briefly been reviewed. We are thinking that this line of study is a new field for the chemical processing of wood with larger future potentials. To achieve the progress in this field, more fundamental and critical studies should be made.

REFERENCES

1. Takase, S. and Shiraishi, N., Function of lignin-carbohydrate complexes within wood cell-walls as a compatibilizer. Abst. Papers Presented at 38th National Meeting, The Japan Wood Res. Soc., Asahikawa, 1988, p.269.

2. Shiraishi, N., Matsunaga, T. and Yokota, T., Thermal softening and melting of esterified wood prepared in an N_2O_4-DMF cellulose solvent medium. J. Appl. Polymer Sci., 1979, **24**, 2361-2368; Shiraishi, N., Plasticization of wood. Mokuzai Gakkaishi, 1986, **32** (10), 755-762.

3. Matsuda, H. and Ueda, M., Preparation and Utilization of Esterified Woods Bearing Carboxyl Groups VIII. Crosslinking of esterified woods with bisepoxide and properties of crosslinked woods. Mokuzai Gakkaishi, 1985, **31** (11), 903-911.

4. Morita, M., Shigematsu, K. and Sakata, I., On physical properties of post-chlorinated cyanoethylated wood. Abst. Papers Presented at 35th National Meeting, The Japan Wood Res. Soc., Tokyo, 1985, p.216.

5. Tsujimoto, N., Yamakoshi, M. and Fukuchi, R., Study on wood-phenol resin fiber. Preprints for Poster Presentation, International Symposium on Wood and Pulping Chemistry, Vancouver, 1985, p.19-20.

6. Shiraishi, N., Tsujimoto, N. and Pu, S., Methods in preparing solutions of lignocellulosic materials. Japan Pat. (Open) 1986-261358.

DISCUSSION

Q: W. G. Glasser (Virginia Tech.)
     (1) If wood dissolves so easily when chemically modified, can lignin be a network polymer ?
     (2) What is the molecular weight of the dissolved wood ?

A: (1) There is a possibility that lignin has not so highly developed three-dimensional structure. Furthermore, during the dissolution, lignin is degraded by phenolysis.

(2) The molecular weights of the components of dissolved wood depend on the methods of dissolution (liquefaction). There are at least three kinds of dissolution of wood. The post-chlorination method gives high molecular weights comparable with natural ones. The liquefaction at high temperature results in low-molecular weight wood components.

Q: E. Sjöström (Helsinki Univ. Tech.)

I would like to ask what was the concentration of wood when it was dissolved. Because the dissolution was made at a high temperature, what was the solubility then at a lower temperature (room temperature) ? Did you obtain a precipitate when lowering the temperature ?

A: When wood liquid is prepared at high temperature or by use of catalyst, wood component becomes lower molecular weight ones than the original. This fact permits concentration as high as 50 wt% or more. In these cases, the liquid system is usually remainable even at room temperature. However, when the liquefied wood components remain to have high molecular weights, lowering of the temperature causes precipitation some times. This occurs when woods are dissolved by the post-chlorination method, and at that time the concentration cannot become higher one.

Q: M. Sakamoto (Tokyo Inst. Tech.)

Would you tell us the yield of carbon fibers from wood-based fibers in comparison with that for cellulosic fibers ?

A: To this question, I asked Mr. Tsujimoto (Oji Paper Co.) to answer. He said the yield is lower than that from pitch fibers.

Q: J. J. Silvy (Ec. Franc. Papeterie)

May we have a comment about the compartment of these dissolved chemically modified wood in respect of the aging that is to say : stability to light and heat in respect of the time ?

A: The dissolved modified wood is stable to light and heat for more than one or two months. However, it is very reactive and if reactive substances are put in, it becomes not stable. For example, if formaldehyde is mixed within wood-phenol liquid and pH is adjusted, resol resin is easily obtainable.

ENGINEERING OF CONTROLLED CELLULOSE-SYNTHETIC POLYMER GRAFT COPOLYMER STRUCTURES

RAMANI NARAYAN
Laboratory of Renewable Resources Engineering,
Purdue University, W.Lafayette, IN 47907, USA

ABSTRACT

Tailor-made cellulose-synthetic polymer graft copolymers with precise control over molecular weight, degree of substitution, backbone-graft linkage, and the overall grafting process, have been synthesized. Crosslinked cellulose graft copolymers with exactly defined polymer chain segments between crosslink points have also been prepared. The graft copolymers exhibit a two-phase morphology and can function effectively as compatibilizers/interfacial agents for alloying/blending cellulosics with synthetic polymers. Incorporation of naturally occurring biopolymers like cellulose into plastics should lead to new type of plastics having the trait of biodegradability.

INTRODUCTION

Considerable research and developmental effort in "macromolecular engineering" of cellulose to produce a composite containing cellulose and grafted synthetic polymer has been expended [1]. The graft polymerization approach to produce a cellulosic molecular composite offers the potential to prepare new types of materials with specific and desirable properties. This is because grafting frequently results in the superposition of properties relating to backbone and side chain [2-5], for example the grafting of 8 to 10 % polyacrylic acid onto high density polyethylene backbone left the melting point and crystallinity unchanged while imparting to it increased modulus and softening point. Grafting is especially useful when the backbone polymer and the side-chain graft are incompatible. Cellulose-synthetic polymer graft copolymers of controlled structure and defined architecture (i.e. precise control over molecular weight and molecular weight distribution, degree of substitution, nature and point of attachment) can function effectively as compatibilizers/interfacial agents for blending/alloying cellulosics with other synthetic polymers. Today, polymer alloys are being successfully used in an increasing number of applications because it provides a means of combining the useful properties of different molecular species to meet new market applications with minimum development cost [6,7]. Block and graft copolymers of the form A-B have been widely used in the polymer industry as "compatibilizers" to improve interfacial adhesion and reduce interfacial tension between A-rich and B-rich phases, resulting in A-B alloys tailored to meet specific needs with unique balance of properties [6,8].

There is, today, an increasing concern over the disposal of persistent, non-degradable plastics as they contribute to a growing share of the general litter problem and the associated solid waste disposal problem. There is also ample evidence that plastic wastes present a hazard to living wild resources, particularly in the marine environment [9]. This has put pressure on federal and state legislatures to mandate plastics degradability — a push that affects an annual 1.8 billion pounds of business, mostly packaging. Incorporation of natural polymers like cellulose in to plastics by alloying or graft polymerization should lead to a new type of plastic having the trait of biodegradability.

Current approaches using radical polymerization methods does not allow for the preparation of controlled cellulose-synthetic polymer graft copolymer structures with defined architecture. Therefore, the potential for creating new cellulosic alloys or graft copolymers with distinctive combination of properties tailored to meet specific applications has not been widely realized. In this paper, we review a new approach developed by us to synthesize cellulose-synthetic polymer graft copolymers of controlled structures with precise control over molecular weight and molecular weight distribution of the graft, degree of substitution, backbone-graft linkage and, overall, better reproducibility of the grafting yields, properties and other features of the graft copolymers.

Current Synthetic Approaches to Cellulosic Graft Copolymers

Nearly all of the grafting methods reported in the literature have involved free radical polymerization under heterogeneous conditions and several extensive reviews have been written on the subject [1, 10-13]. This essentially involves generating radical sites on the cellulose backbone using radiation or chemical methods which initiates polymerization of the added vinyl monomer. Unfortunately, homopolymerization of the vinyl monomer also occurs. The homopolymer gets embedded into the cellulose/starch matrix and is difficult to remove. Further, one has no control over the position where the reactive sites are initiated on the backbone polymer. As a result, there is no control nor the ability to change the nature of the backbone-graft linkage or the points of attachment to the backbone. There is no control over the molecular weight of the synthetic polymer branch to be grafted onto the backbone nor is there the ability to graft narrow molecular weight (monodisperse) branches. Cross linking reactions can and do occur under these conditions. Thus, there is very little control over the grafting process and no ability to engineer precise tailor-made cellulose-synthetic polymer graft copolymer structures.

In a different approach alkali metal cellulosates have been used as initiators for the anionic graft copolymerization of vinyl monomers onto cellulose [14-15]. Although this method seems more versatile, it does not allow adequate structural control and polydisperse products are often produced. An excellent review article by Stannett [16] discusses problems and potential of cellulosic graft copolymers for industrial applications.

RESULTS & DISCUSSIONS

New Synthetic Approach

We have developed a new approach that allows us to engineer cellulose-synthetic polymer graft copolymers with precise control over molecular weight and molecular weight distribution, degree of substitution, backbone-graft linkage and with reproducible grafting yields and properties of the graft copolymers. The approach involves:

1) Preparation of monodisperse, desired molecular weight synthetic polymer anion by anionic polymerization. Anionic polymerization methods are widely used in the synthesis of well defined polymer structures with control over molecular weight and having a narrow molecular weight distribution. The anionic polymerization system offers unique opportunities for the synthesis of block and graft copolymers of known architecture in regard to both composition and chain length and the potential to prepare well defined homopolymers with functionalized end groups. In the present case, polystyryl carbanion of desired molecular weight and narrow molecular weight distribution was

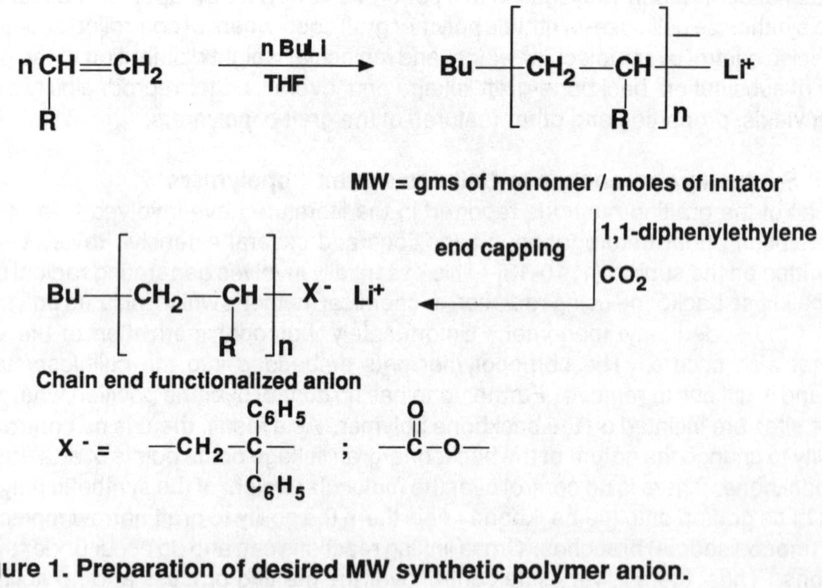

$$MW = \text{gms of monomer / moles of initator}$$

Chain end functionalized anion

Figure 1. Preparation of desired MW synthetic polymer anion.

prepared by anionic polymerization methods using n-butyllithium in THF. The carbanion (living polymer) was capped with 1,1-diphenylethylene or CO_2 to provide the 1,1-diphenylethane carbanion or the polystyrylcarboxylate anion respectively. The end capping of the carbanion with the 1,1-diphenylethylene or CO_2 was necessary to reduce the basicity of the polystyryl carbanion. Otherwise, the highly basic polystyryl carbanion readily abstracts a glycosidic proton with concurrent bond cleavage resulting in degradation of the cellulosic chain [17].

2). Functionalization of the cellulose backbone. Sulfonic ester groups (mesylate or tosylate) was introduced onto a acetylated cellulose backbone. The sulfonic ester groups function as good leaving groups in nucleophilic displacement reactions with the

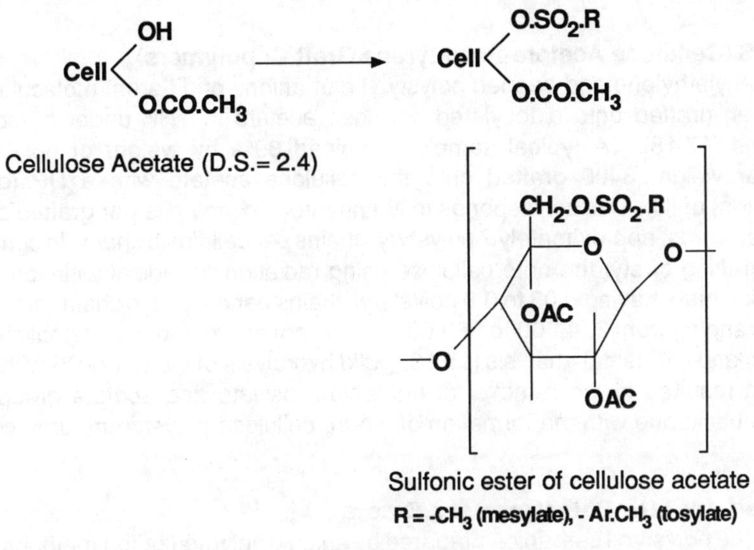

Cell $\big\langle$ $\begin{array}{l} OH \\ O.CO.CH_3 \end{array}$ $\longrightarrow$ Cell $\big\langle$ $\begin{array}{l} O.SO_2.R \\ O.CO.CH_3 \end{array}$

Cellulose Acetate (D.S.= 2.4)

Sulfonic ester of cellulose acetate

$R = -CH_3$ (mesylate), $- Ar.CH_3$ (tosylate)

Figure 2. Functionalization of the cellulose backbone.

synthetic polymer anions generated by anionic polymerization approaches as outlined in (I). Acetylated cellulose was used to obtain a organic solvent (THF) soluble product and to impart thermal plasticity to the final graft copolymer product.

3). Coupling of the synthetic polymer anion with the cellulose acetate sulfonic ester. The synthetic polymer anion from step (1) was reacted with the mesylated or

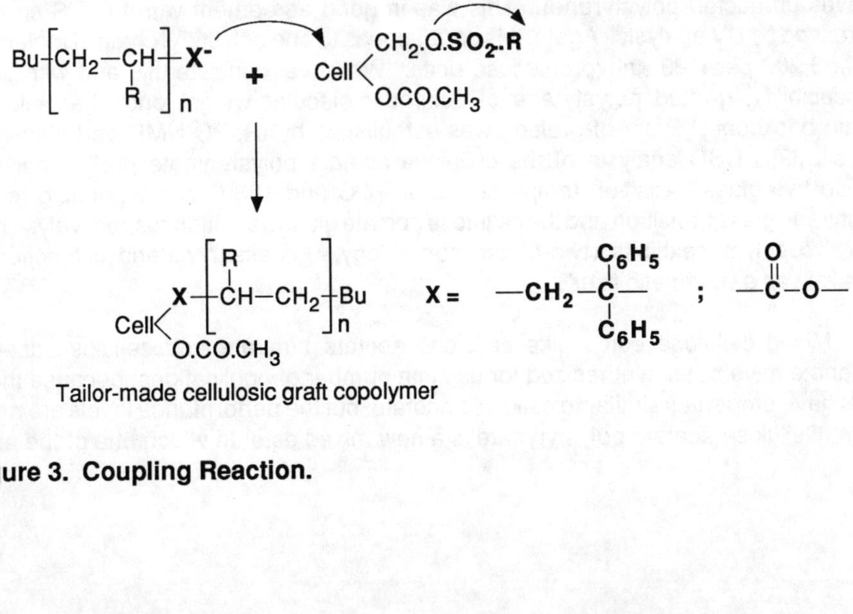

$Bu \text{-}\Big[CH_2\text{-}CH\Big]_n\text{-}X^-$ (with R) $+$ Cell $\big\langle$ $\begin{array}{l} CH_2.O.SO_2.R \\ O.CO.CH_3 \end{array}$

Cell $\big\langle$ $\begin{array}{l} X\text{-}\Big[CH\text{-}CH_2\Big]_n\text{-}Bu \text{ (with R)} \\ O.CO.CH_3 \end{array}$ $X = -CH_2\text{-}\underset{C_6H_5}{\overset{C_6H_5}{C}}-$; $-\overset{O}{\overset{\|}{C}}\text{-}O-$

Tailor-made cellulosic graft copolymer

Figure 3. Coupling Reaction.

tosylated cellulose acetate under homogeneous conditions. The carbanion displaces the sulfonic ester group in a typical SN2-type nucleophilic displacement reaction with the formation of controlled, well defined cellulose-synthetic polymer graft copolymer structures.

CA-G-PS (Cellulose Acetate-Polystyrene Graft Copolymers)

1,1-diphenylethylene end-capped polystyryl carbanions of different molecular weights have been grafted onto a tosylated cellulose acetate in THF under homogeneous conditions [17,18]. A typical sample contained 39% by weight of polystyrene of molecular weight 3400 grafted onto the cellulose acetate with a DP (degree of substitution) of ll0. This corresponds to l6 anhydroglucose units per grafted side chain or in other words, approximately 7 polystyryl chains per cellulosic chain. In contrast, free radical grafting of styrene onto cellulose using radiation or radical initiation produces graft copolymers having 0.03 to 0.8 polystyryl chains per cellulosic chain with molecular weights ranging from 354,000 to 960,000 [l6]. Proof of grafting was established by IR, solubility, and elemental analysis [17,18]. Mild hydrolysis of the CA-g-PS with aqueous ammonia resulted in the removal of unreacted tosylate and acetate groups on the cellulose backbone with the formation of a pure cellulose-polystyrene graft copolymer [18].

Cellulose Acetate Polystyrynate Esters

Capping the polystyryl carbanion prepared by anionic polymerization methods with CO_2 resulted in the formation of a polystyrylcarboxylate anion (Figure 1). This anion is not sufficiently reactive to displace acetate groups from cellulose acetate. It is, however, sufficiently nucleophilic to displace better leaving groups like mesylate groups from mesylated cellulose acetate with concomitant formation of an ester linkage [l9]. A further advantage of the direct use of polystyrylcarboxylate anion over the polystyryl carbanions is that water does not interfere with the grafting reaction. Table 1 shows the results of grafting experiments between polystyrylcarboxylate anion and the mesylated cellulose acetate at 75°C for 20 hours. The polystyrene content of the graft copolymer was around 55-57% based on the weight increase after toluene extraction (toluene extraction removes unreacted polystyrene). This was in good agreement with the PS content determined by UV analysis. Again, this corresponds to one polystyryl chain of molecular weight 6,200 per l7-l9 anhydroglucose units. We have, consistently, and with good reproducibility, grafted polystyrene of desired molecular weight onto the cellulose acetate backbone. Proof of grafting was established by IR, ^{13}C NMR and elemental analysis [l9]. DSC analysis of the cellulose acetate polystyrynate graft copolymer showed two glass transition temperatures at 97°C and 198°C corresponding to the polystyrene glass transition and the cellulose acetate glass transition respectively. Thus, the graft copolymer exhibits a two-phase morphology, a necessary criteria for functioning effectively as a "compatibilizer".

Mixed cellulose esters like cellulose acetate butyrate and cellulose acetate propionate have been synthesized for use in a number of applications, because these esters have properties similar to cellulose acetate, but the performance levels are much better. Cellulose acetate polystyrynate is a new mixed ester in which one of the ester

Table 1
Results of Grafting of Polystyrene
Monocarboxylate onto Mesylated Cellulose Acetate
at 75°C, 20 Hours

Product No.	Solvent	Grafting[a] Yield, %	PS Content[b,c] (Wt. %)	PS Content by UV (Wt. %)	AGU[d] per PS Chain
3a	DMF	68	57.6	58.2	17.0
3b	DMSO/THF60	60	54.5	59.9	19.3

[a]Percent by weight of the PS grafted.
[b]Determined from the weight increase.
[c]The molecular weight of the monodisperse polystyryl monocarboxylate is 6200.
groups is a thermoplastic polymer chain and should have unique properties. During processing, the thermoplastic polystyrene graft chain can function as an internal plasticizer eliminating the use of plasticizer additives with its attendant problems.

Crosslinked Cellulose Graft Copolymers With Exactly Defined Polymer Chain Segments Between Crosslink Points
If in the preparation of the synthetic polymer anion, the butyllithium initiator was replaced by sodium naphthalene, a difunctional carboxylate anion would be formed. Since both

ends of the synthetic polymer chain can potentially react with the mesylate groups on the

Table 2
Results of Grafting Polystyryl
Dicarboxylate Anion with Mesylated Cellulose Acetate
at 75°C, 20 Hours

Product No.	Solvent	Grafting[a] Yield, %	PS Content[b,c] (Wt. %)	AGU[d] per PS Chain
4a	DMF	90.5	64.4	22.5
4b	DMSO/THF	88.5	63.9	23.0

[a]Percent by weight of the PS grafted.
[b]Determined from the weight increase.
[c]The molecular weight of the monodisperse polystyryl dicarboxylate anion is 10,900

cellulose backbone crosslinking is expected to occur. Indeed, the reaction of pol-ystyryldicarboxylate anion with mesylated cellulose acetate resulted in the formation of a solid gel, indicative of crosslinking [19]. The results are shown in Table 2. Based on the polystyrene content of the graft copolymer product and the molecular weight (10,900), there is one polystyryl crosslink for every 23 anhydroglucose units. While crosslinking of cellulosics, especially in the textile industry, has been done using, monomers, there has been to our knowledge, no reports of crosslinking cellulose acetate or other cellulose derivatives with synthetic polymer chains. Thus, this is the first synthesis of a crosslinked cellulosic graft copolymer with exactly defined polymer chain segments between the crosslink points.

In the crosslinking reaction the key step is the nucleophilic displacement of a mesylate group by a synthetic polymer carboxylate anion. Therefore, carboxy termi-nated synthetic polymer or a synthetic polymer carrying carboxy precursor group can be grafted onto mesylated cellulose acetate. Thus, partially hydrolyzed poly(methyl-methacrylate) was successfully grafted onto mesylated cellulose acetate in excellent yields by nucleophilic displacement of mesylate groups in less than 16 minutes at 75°C resulting in a polymethylmethacrylate crosslinked cellulose acetate graft copolymer [20]. A commercially available polyamide resin (Aldrich) formed by condensation of polyam-ine with dibasic carboxylic acids produced from unsaturated fatty acids was also grafted onto a mesylated cellulose acetate resulting in a graft copolymer of polyamide with cellulose acetate [21].

REFERENCES

[1].   A. Hebish, and J. T. Guthrie, *The Chemistry and Technology of Cellulosic Copolymers,* Springer-Verlag, New York (1981).

[2].   H. P. Hopfenberg, V. Stannett, F. Kumura-Yeh, and P. T. Rigney, *Appl. Polym. Symp. 13,* 139. (1970).

3].   J. K. Rieke, and G. M. Hart, *J. Polym. Sci., Part C,* 1, 117 (1963).

[4].   J. K. Rieke, and G. M. Hart, *J. Polym. Sci., Part C,* 4, 589 (1964).

117

[5]   V. Stannett, in B*lock and Graft Copolymers*, J. J. Burke, and V. Weist, eds., Syracuse University Press, Syracuse, 1973, p. 281.
[6]   Modern Plastics, p. 62, October 1988
[7]   V. Wigotsky, *Plastics Eng.*, XLIV (11), 25, November 1988
[8].   D. R. Paul, and S. Newman, eds., *Polymer Blends*, 182, Academic Press, New York,1978.
[9].   A. L. Andrady, *Proceedings of the SPI Symposium on Biodegradable Plastics*, June 10, 1987, The Society of Plastics Industry, USA
[10]. J. C. Arthur, Jr., *Appl. Polym. Symp. 36*, 201, (1981).
[11]. O. Y. Mansour, and A. Nagaty, *Prog. Polym. Sci. 11*, 91, (1985).
[12]. S. N. Bhattcharyya, and D. Maldas, *Prog. Polym. Sci. 10*, 171, (1984).
[13]. B. R. Morin, I. P. Breusova, and Z. A. Rogovin, *Adv. Polym. Sci., 42*, 139, (1982).
[14]. M. Tahan, B. Yom-Tov, and A. Zilkha, *Europ. Polym. J., 5*, 499 (1969)
[15]. Y. Avny and L. Rebenfield, *Textile Res. J., 38*, 684 (1968)
[16]. V. Stannett, V. *ACS Symp. Ser., 187*, 1., (1982)
[17]. R. Narayan, and M. Shay, M. in *Renewable Resource Materials, New Polymer Sources*, C. E. Carraher, Jr. and L. H. Sperling, eds., Plenum; *Polym. Sci. Tech. 33*, 137, (1986)..
[18]. R. Narayan, and M Shay, M. in *Recent Advances in Anionic Polymerization*, T. E. Hogen-Esch, and J. Smid, eds., Elsevier, New York, (1987) pp. 441-450.
[19]. C. J. Biermann, J. B. Chung, and R. Narayan, R., *Macromolecules 20*, 954, (1987).
[20]. C. J. Biermann, and R. Narayan, *Polymer 28*, 2176, (1987).
[21].   C. J. Biermann, and R. Narayan, *Forest Prod. J., 38(1)*, 27 (1988).

ACKNOWLEDGEMENTS
This material is based on work supported by the National Science Foundation under Grant No. CBT-8502498

DISCUSSION

Q: Y. Tezuka (Tech. Univ. Nagaoka)
    During the coupling reaction between cellulose derivative and synthetic polymer with such functionality, the phase separation may occur due to the incompatibility of the two polymers. Is it not the case in your reaction system ?

A: Phase separation may occur, but by using dilute solutions and stirring well, and running the reaction in a solvent in which all the reactants are soluble, very good reactivity between the synthetic polymer anion and the mesylated/tosylated cellulose acetate.

Q: P. K. Chatterjee (J & J Absorbent Tech.)
    The cellulose graft copolymer approach which you describ-

ed will allow to prepare film, block polymer, powder as end
products but if we are interested in preparing graft copolymer mainly the cellulose fibrous structure intact, how
shall we do it ? The key point here is to maintain the cellulose fiber structure intact.

A: It should be possible, although the reaction (in all
probability) would be heterogeneous reaction. (1) Mesylate
groups would have to be introduced on the fiber (it does not
have to be completely mesylated). (2) The designed MW monodisperse synthetic polymer anion to be prepared by anionic
polymerization and (3) coupling reaction of the mesylated
cellulose fiber with the synthetic polymer anion. Indeed we
have reacted mesylated wood fiber with polystyryl carboxylate
anion.

Q: C. A. Steiner (City Univ. NY)
   Dr. Narayan, have you performed any analysis to determine
how the grafts are distributed along the backbone ? Are they
randomly distributed or do they tend cluster together ?

A: We have not done any detailed analysis. However, the
grafting takes place exclusively on the C-6 position of the
anhydroglucose unit and I believe would be randomly distributed on the cellulose backbone especially in view of the high
degree of substitution we obtain. - 1 PS chain per 17 anhydroglucose units. -

ROUND TABLE I-2   <Hi-Tech>

"ORGANIZED MOLECULAR ASSEMBLIES"

Chairman: P. Zugenmaier
           (Tech. Univ. Clausthal)

INTRODUCTORY REMARKS

given by P. Zugenmaier

Mr. Tatsuo Tanabe, president of Nisshinbo Industries
stated in his message to the participants of this conference:
"Last year, when we celebrated our 80th anniversary, we
established a new vision for the future, and we made a new
start toward the 21st century". A year ago, I read in "The
Japan Times" a critical article by Ichio Takenaka about the
role of Japan in the 21st century and about changes to be
made to cope with the markets: "Japan has a key role to play
in developing an interdependent, affluent Asia: We must provide the institutional frame work for basic research and
applied technology. But we lack the technostructure - university facilities and research institutes - that has given the
U.S. a lead in advanced research and development."

This conference organized by an industrial company already provides a clear vision about the application and basic
research on cellulosics and other polysaccharides and it
brings together researchers from all over the world to
express their ideas on the utilization of cellulosics in the
near future. One of the topics of this meeting will be "Organized Molecular Assemblies". Knowledge of organized molecular
assemblies is of vital interest for almost all applications
of various kinds of polymers in the solid state, liquid crystalline state and in solution where the formation of a certain structure is a requisite for many uses. Cellulosic material provides an additional feature, the chirality, which is
a necessity in many applications. Two papers will be presented in this Hi-Tech section which deal exclusively with the
chirality of the cellulose molecule, three papers present
special structures and assemblies.

Cellulose derivatives may serve as sorbants for separation of enantiomers in high-performance liquid chromatography (HPLC). The compounds resolved include many drugs. It is
of eminent importance to recognize the active compound and
test the inactive one for side effects. Great progress is
achieved in this field and many polysaccharide derivatives,
including a respectable number of polysaccharide phenylcarbamate deriv-atives, have been used successfully as stationary
phase in chromatography columns. However, very little is
known about the interaction mechanism, the so called chiral
recognition of the sorbants. This knowledge should lead to
improvements in separation and better selections of the systems.

The chirality of the cellulosic molecule is also necessary for the formation of lyotropic cholesteric liquid crystals with some special applications appearing. Chiroptical
filters may be prepared with high selectivity in a very
narrow wavelength band. This fourth state of matter can also
be used as an analytical tool e.g. for a determination of
molecular mass of cellulose, and it has been proposed, that
high modulus fibers may be drawn from lyotropic compensated
cholesteric phases. Very little is known about the structure

forming interactions of the cellulosic molecule and the solvent which lead to a supermolecular helicoidal arrangement with right- or left-handed twist.

Quite a number of applications are appearing by the basic study of polyelectrolyte complex (symplex) formation through interactions between polyanions and polycations of various kinds including anionic cellulose derivatives. The scientific value of such an investigation lies in an understanding of the principles of the formation of higher organized polymer structures. Promising fields of applications are the encapsulation of living cells and cell assemblies at quasi-phisiological conditions as well as promotion of flocculation of dispersed or dissolved compounds from effluents e.g. the binding of lignosulphonate, as well as coating.

A further field from the practical and academic point of view is represented by Langmuir-Blodgett films of polymers. It is possible to create well ordered uniform films of molecular dimensions by this method and to study structure-properties relationship which might not be available by other means. Since cellulose is a hydrophilic, linear compound, it can be converted into amphiphilic derivatives, necessary for monolayer formation by the Langmuir-Blodgett technique, by introducing hydrophobic substituents. Excellent films have been formed by cellulose esters with alkyl side groups and examined by surface pressure-area isotherms and electron microscopy. This seems to be the first detailed study on the fine structure of cellulose derivative monolayers and may open a promising field for further investigations.

Tailored cellulose derivatives play an important role as surfactants in the presence of dispersed solids and surfactant micelles. The grafted side chains on hydroxyethyl cellulose have the ability to associate with the dispersed phase forming distinct molecular assemblies. Altered rheological and filtration properties result for the system. The concentration and structure of the surfactant as well as the total concentration of the polymer determine the bulk property of the system, e.g. an increase of viscosity is expected before the critical micelle concentration is reached, and the viscosity decreases near the critical micelle concentration. This very unusual behavior of solutions opens a new and interesting field, and many applications can be visualized in the near future.

The organizers of this conference have actually achieved a high ranking goal in bringing together scientist from fundamental and applied research, from which the development of high tech products will certainly benefit. Having this excellent cellulose conference in mind, I cannot agree with the above quoted newspaper article that "Japanese research and development centers ... lack vital input from university programs in basic and applied research, a requisite of long-term technical innovation".

Let me close this introduction for the High-Tech session "Organized Molecular Assemblies" with a personal remark. I have often wondered and I am puzzled how nature forms stable structures e.g. polymeric crystals and molecular assemblies so quickly, yet sometimes these structures cannot be solved with powerful and fast computers even after long computations.

During a stay in Kyoto, I have visited the magnificent gardens of Shugakuin Imperial Villa. I was overwhelmed by the amazing beauty of the gardens and teahouses constructed in clear and simple lines. May we learn from the architects of these gardens and teahouses and introduce clear and simple ideas to solve the complex problems.

CHOLESTERIC STRUCTURE IN CELLULOSIC FILMS

DEREK G. GRAY
Department of Chemistry, McGill University
3420 University St., Montreal, Quebec, Canada   H3A 2A7

ABSTRACT

Films of cellulose and cellulose derivatives may be prepared with molecular orientations similar to the helicoidal supramolecular arrangement characteristic of cholesteric liquid crystals. Films of (hydroxypropyl)cellulose cast from aqueous liquid crystalline solutions reflect
circularly-polarized UV light. On heating, the reflection wavelength
increases irreversibly to the visible range. Measurement of the circular reflectivity (apparent circular dichroism) of the films indicates
that the polymer molecules are in a right-handed helicoidal arrangement.   An artifact due to residual shear-induced birefringence may
reverse the apparent handedness of these films. Cholesteric order may
also be inferred from the observation of induced circular dichroism
from Congo red dye in specially-prepared regenerated cellulose films.
Transmission electron microscopy of film cross-sections suggests that
the cholesteric pitch in films of cellulose acetate and regenerated
cellulose is much smaller than the wavelength of visible light.

Considerable effort has been devoted to the preparation of free-stan-

ding polymeric films exhibiting cholesteric order. Different routes

have been studied, including (i) synthesis of thermotropic polymer

liquid crystals with cross-linkable side chains (1), (ii) incorporation

of cholesteryl derivatives as side chains in acrylate (2) and siloxane

polymers (3), (iii) bulk polymerization of polymeric lyotropic meso-

phases in vinyl solvents (4,5), (iv) quenching of copolyesters from

their thermotropic range to room temperature (6), and (v) casting from

a lyotropic liquid crystalline solution (7,8). In this paper, some

results on the formation of cholesteric arrangements in films of cellu-

lose and cellulose derivatives from this last route will be summarized.

Detection of cholesteric order in films is straightforward if the pitch

of the cholesteric struture is of the magnitude of the wavelength of visible light. Reflection of circularly polarized light by the film is normally a clear indication of cholesteric structure in the film.

(Hydroxypropyl)cellulose films cast from water: (Hydroxypropyl)-cellulose films cast from water solutions assume a cholesteric arrangement, with pitches in the range 100-200 nm, depending on the casting conditions (9). The reflection wavelength moves into the visible range upon heating between 130 and 170°C, the thermotropic order being characterized by a positive temperature dependence of the cholesteric pitch. This process is irreversible; the pitch values decrease only slightly on cooling. The reflection band forms rapidly on heating, confirming that the cholesteric order was already present in the cast films at room temperature. Figure 1 shows the reflectivity of a water-cast film on stepping the temperature from ambient to 160°C. The cholesteric band appears immediately on heating, but drifts with time to longer wavelengths, becoming essentially stable after three hours. The final reflection wavelength also increases with temperature in the range 140-170°C. The cholesteric properties of (hydroxypropyl)cellu-lose films are therefore metastable. The persistence of cholesteric order in the cast films at room temperature and the behaviour of the optical properties on heating are consistent with dynamic mechanical measurements, which reveal an increase in the molecular mobility above 120°C (9).

Effect of residual linear orientation: The interpretation of the cholesteric structure of films based on measurements of circular reflectivity (apparent circular dichroism) is subject to some arte-facts. Perhaps the most serious is the effect of residual linear orientation of the film on the reflection band. HPC films cast from an aqueous mesophase show the negative circular reflection band charac-teristic of right-handed helicoidal structures (Figure 1). However, films cast from mesophase solutions that have been sheared and then dried before relaxation to the equilibrium cholesteric structure is complete show a positive circular reflection band (10). At room tem-perature this reflection band is situated below 300 nm but, as is the case for cholesteric HPC films, it moves to longer wavelengths on heating. Spectra recorded for this type of samples when heated at 150°C are shown in Figure 2. This behaviour is very similar to that

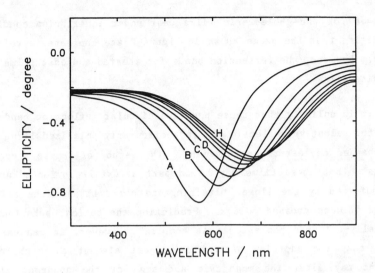

Figure 1. Circular reflectivity of a water-cast HPC film at 160°C. The sample was placed in the preheated hot stage. Spectra were recorded 2 min after heating (A) and then at 15 min intervals (B-H). (From reference 9, with permission.)

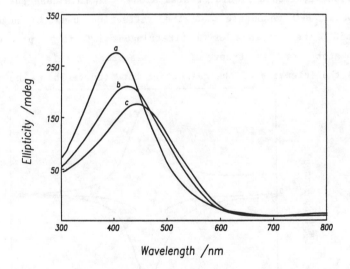

Figure 2. Circular reflectivity spectrum of an HPC film cast from aqueous mesophase solution under shear recorded 2 min (a), 12 min (b) and 40 min (c) after heating from room temperature to 150°C. (From reference 10, with permission.)

observed for the cholesteric films cast under equilibrium conditions, indicating that the peaks shown in Figure 2 are cholesteric reflection bands. However, the reflection bands for sheared and cast films are of opposite signs.

It is unlikely that these positive circular reflection bands arise from the selective reflection of left-circularly polarized light from a left-handed cholesteric arrangement. The method of sample preparation suggests that these films may be in part uniaxially oriented and this is confirmed by the linear birefringence observed when the films are viewed between crossed polars. Predicting the optical behaviour of a partially-oriented cholesteric is complex. However, it can easily be shown that the addition of a birefringent element to a cholesteric element may alter the magnitude and sign of the apparent circular reflectivity (10). By combining thin (< 5 μm) oriented films of HPC with aqueous cholesteric layers of the same polymer, the sign of the optical rotation can be reversed. CR spectra for such a combination of samples are shown in Figure 3. The bottom curve (3a) is the reflection band of the lyotropic liquid crystal alone. In this case the sample is an aqueous HPC mesophase and the reflection band is negative as expected. The presence of a birefringent HPC film, placed in the spectrometer beam in front of the mesophase sample, significantly reduces the intensity of the reflection band (curve 3b). The introduc-

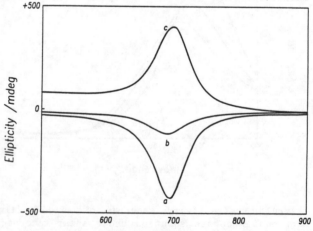

Figure 3. Circular reflectivity spectrum of an aqueous HPC mesophase alone in the spectrometer beam (a), with one oriented film in beam (b) and with two oriented films in the beam (c). (From reference 10, with permission.)

tion of a second piece of the same film affects the spectrum differently depending on the orientation of its shear direction relative of that of the first. Two oriented films placed so that the shear direction of one is perpendicular to that of the other have no effect on the mesophase spectrum, while the introduction of a second oriented film parallel to the first reverses the sign of the peak (curve 3c). If the samples are placed so that the spectrometer beam first passes through the mesophase sample and then the oriented film, the reflection spectrum is unchanged from that recorded for the mesophase alone (10).

Thus, the magnitude and handedness of the reflection band from a cholesteric HPC mesophase may be modified and even reversed by passage through an oriented HPC film acting as a waveplate. It thus seems plausible that this combination of cholesteric structure and linear orientation, induced by shear in films of HPC, results in the apparent reversal of handedness illustrated in Figure 2.

Cholesteric films of cellulose: If cellulose derivatives form films with helicoidal supramolecular structure, does cellulose itself form similar structures? Unsubstituted cellulose has been reported to form a lyotropic mesophase in LiCl/N,N-dimethylacetamide (11,12) and in a mixture of N-methyl-morpholine N-oxide and water (13). Although a high optical rotation was measured for the anisotropic phase (11), no cholesteric reflection has been observed for cellulose. Assuming that anisotropic solutions of cellulose can be prepared, the problem remains of making cellulose films with cholesteric structure, and demonstrating that such structure exists.

When cellulose films regenerated from LiCl/DMAC solution are dyed with Congo red, the dye shows induced optical activity (14). The induced optical activity is absent for dyed films which are prepared under conditions which preclude or disrupt molecular ordering.

When examined with a polarizing microscope, cellulose acetate films cast from mesophase solution in trifluoroacetic acid exhibit the parabolic focal conic texture typical of cellulose-based lyotropic mesophases. This texture is unchanged by removal of acetate groups during the regeneration of cellulose in aqueous ammonia (14). The absence of changes in morphology despite a major chemical modification, while somewhat surprising, suggests that the molecular order present in the cellulose acetate films is preserved during deacetylation. Both isotropic and cholesteric cellulose films can therefore be prepared from the corresponding cellulose acetate films.

The induced circular dichroism spectrum recorded for a cellulose film regenerated from cholesteric cellulose acetate and then dyed with Congo red is similar to that recorded for cellulose films regenerated from LiCl/DMAC solution (14). If it is assumed that the former represents the type of optical activity induced in achiral dyes oriented in a cholesteric matrix, this observation implies that cellulose adopts cholesteric order on slow precipitation from solution. The film regenerated from an isotropic cellulose acetate film showed no induced dichroism. The induced circular dichroism discussed above indicates that the cellulose films prepared from lithium chloride/dimethyl acetamide solution and by deacetylation of cellulose acetate or triacetate films cast from liquid crystalline solutions in trifluoroacetic acid display a cholesteric structure.

Transmission electron microscopy of cholesteric structure in films of cellulose acetate and cellulose: The results from induced circular dichroism (14) do not indicate the magnitude or handedness of the pitch of the cholesteric structure. No reflection of visible light was observed for the cellulose or cellulose acetate films. It seemed likely that the pitch was shorter than the wavelength of visible light, and therefore further evidence for a helicoidal structure was sought by transmission electron microscopy.

The high contrast images in Figure 4 are examples obtained on cross-sections of a cellulose acetate film cast from trifluoroacetic acid (15). A clear and well defined lamellation is observed, with a periodicity ranging from 20 to 35 nm in the zones where the contrast is highest. These zones probably correspond to areas where the lamellae are almost perpendicular to the plane of observation. Other regions with lower contrast and broader spacings may correspond to areas where the lamellae are viewed obliquely. When deacetylated, this film retained the fingerprint texture. The cellulose film obtained by

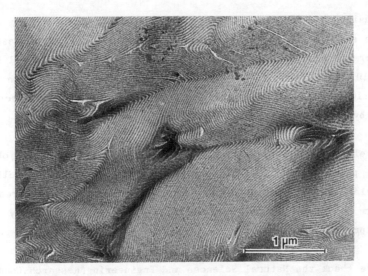

Figure 4. Transmission electron micrograph of cross-section of cellulose triacetate film prepared from cholesteric mesophase in
trifluoroacetic acid. (From reference 15, with permission.)

precipitation from LiCl/DMAC also shows a lamellation which, however,
is less well defined. The cellulose triacetate films are amorphous but
relatively well-ordered orientationally in a helicoidal cholesteric
arrangement; dislocations and focal lines are clearly identifiable.
Electron micrographs of a cellulose film made from an isotropic solution of cellulose acetate showed a complete absence of the periodic
structure. The fingerprint texture evidently reflects periodicities in
electron density in the thin sections that are related to the helicoidal supramolecular arrangement characteristic of cholesteric liquid
crystals, and indicates that this arrangement is present not only in
cellulose regenerated from cholesteric cellulose acetate films, but
also in cellulose precipitated directly from LiCl/DMAC (15).

Patterns similar to the electron micrographs in Figure 4 were
first observed in optical micrographs of concentrated polypeptide solutions (16), and subsequently in solutions of cellulose derivatives
(5,17). The fingerprint patterns result from the periodicity in
refractive index normal to the axis of the helicoidal cholesteric
arrangement. (However, for fluid liquid crystalline layers between
glass plates, the periodicity in the spacing of the fingerprint lines

is usually uniform, and does not show the variations in spacing evident in the sections of solid film shown here.) Thus cellulosic mesophases can form twisted helicoidal structures with pitches ranging all the way from 10 micrometers (observed by optical microscopy) through to the very tightly twisted arrangements with pitches of 40 nm observed in films by electron microscopy.

Helicoidal arrangements of cellulose fibrils have been observed in the plant cell wall (18,19) by electron microscopy. The helicoidal molecular arrangements observed here in synthetically prepared films confirm the strong tendency for cellulose to form spontaneously twisted structures in vivo and in vitro.

We thank the Natural Sciences and Engineering Research Council of Canada for support.

REFERENCES

1. Bhadani, S.N. and Gray, D.G., Mol. Cryst. Liq. Cryst., 1984, 1 0 2 , 255.

2. Shannon, P., Macromolecules, 1984, 17, 1873.

3. Finkelmann, H. and Rehage, G., Makromol. Chem. Rapid Commun., 1980, 1, 733.

4. Nishio, Y., Yamane, T. and Takahashi, T., J. Polym. Sci. Polym. Phys. Ed., 1985, 23, 1043.

5. Tsutsui, T. and Tanaka, R., Polymer, 1980, 21, 1351.

6. Krigbaum, W.R., Ciferri, A., Asrar, J. and Toriumi, H., Mol. Cryst. Liq. Cryst., 1981, 76, 79.

7. Ito, K., Kajiyama, T. and Takayanagi, M., Polym. J. (Tokyo), 1977, 9, 355.

8. Watanabe, J., Naka, M., Watanabe, K. and Uematsu, I., J. Polym. Sci. part A-2, 1969, 7, 1197.

9. Charlet, G. and Gray, D.G., Macromolecules, 1987, 20, 33.

10. Ritcey, A.M., Charlet, G. and Gray, D.G., Can. J. Chem., in press.

11. Conio, G., Corazza, P., Bianchi, E., Tealdi, A. and Ciferri, A., J. Polym. Sci.: Polym. Lett. Ed., 1984, 22, 273.

12. McCormick, C.L., Callais, P.A. and Hutchinson, B.H., Polym. Prepr., 1983, 24(2), 271.

13. Chanzy, H. and Peguy, A., J. Polym. Sci: Polym. Phys. Ed., 1980, 18, 1137.

14. Ritcey, A.M. and Gray, D.G., Biopolymers, 1988, in press.

15. Giasson, J., Revol, J.-F., Ritcey, A.M. and Gray, D.G., Biopolymers, 1988, in press.

16. Robinson, C., Faraday Soc. Trans., 1956, 52, 571.

17. Bheda, J., Fellers, J.F. and White, J.L., Colloïd Polym. Sci., 1980, 258, 1335.

18. Neville, A.C. and Levy, S., Planta, 1984, 162, 370.

19. Vian, B., Reis, D., Mosiniak, M. and Roland, J.C., Protoplasma, 1986, 131, 185.

DISCUSSION

Q: P. Zugenmaier (Tech. Univ. Clausthal)
How do you orient your sample ? Can you comment on sample thickness ?

A: No special methods were used to orient these samples. In our experience, the thinner samples form the planar texture. Thicker samples tend to twist and form focal conic textures.

Q: Y. Nishio (Fukui Univ.)
You have shown an EM photograph of a cellulose film prepared from solution in DMAc-LiCl. How was the polymer concentration of the original solution (CELL/DMAc-LiCl), i.e. from anisotropic solution or isotropic one ?

A: The cellulose thin film was slowly precipitated from a dilute (0.75% cellulose) isotropic solution in DMAc/LiCl by exposure to water vapor.

Q: J. Watanabe (Tokyo Inst. Tech.)
You showed the TEM photographs taken for the thin microtomed films cut out of the cholesteric film. In those films, the black and white lines can be observed with a well contrast. I easily understand that such black and white lines can be attributable to the cholesteric repeating, but conversely, I have a question why the black and white lines with a well contrast are produced by the cholesteric helical structures.

A: We are not sure at present what causes the electron beam scattering. It may be due to the periodicity in the mechanical properties of the film, as revealed by the sectioning procedure.

NEW CELLULOSIC SURFACTANTS

Carol A. Steiner
Dept. of Chemical Engineering, The City College of CUNY, NY, NY 10031

Robert A. Gelman
Research Center, Hercules, Inc., Wilmington, DE 19894

ABSTRACT

Hydrophobically modified hydroxyethyl cellulose (HMHEC) is a novel
surface-active cellulose ether which associates with the suspended
phase in dispersions and surfactant solutions, giving rise to unusual
rheological and filtration properties. In addition this polymer can
be made to self-associate into networks in a controllable manner
depending on the composition of the solvent. This unique behavior
makes this polymer suitable in applications ranging from cosmetics
formulations to latex paints.

INTRODUCTION

Cellulose in its native form is not soluble in water. It has the
capacity to be rendered water-soluble by chemical reaction of its
hydroxyl groups with hydrophilic substituents. In this manner
polymers such as carboxymethyl cellulose (CMC) and hydroxyethyl
cellulose (HEC) are produced. These water-soluble cellulose ethers
may be further modified by reaction with alkyl compounds, forming
graft copolymers whose backbones are water-soluble and whose side
chains tend to aggregate, surfactant-like, in aqueous solutions.
Obviously, the rheological properties and surface activity of the
resulting materials will be dramatically different from those of the
unmodified (water-soluble) polymers.

One such cellulosic surfactant which has been investigated
extensively [1-4] is hydrophobically modified hydroxyethyl cellulose
(HMHEC) (trade name Natrosol Plus[R], Aqualon Co., Wilmington, Del.).
Its backbone is hydroxyethyl cellulose (HEC) (trade name Natrosol[R],
Aqualon Co.), a water-soluble cellulose ether with ethylene oxide (EO)
as the hydrophilic substituent. Alkyl chains 8 to 12 carbons long are
grafted to the polymer via ether linkages with the EO groups.
Typically a hydrophobe level of >0.4% (by weight) alkyl chains is
sufficient to render the polymer insoluble in water. However even at
>0.4% hydrophobes HMHEC is soluble in surfactant solutions and in
mixed solvents such as alcohol and water, where hydrophobic
interactions among the side chains are disrupted by the solvent. In
this paper we will present results showing the behavior of HMHEC in
dispersions and surfactant solutions.

RESULTS AND DISCUSSION

HMHEC is well suited for use in coatings or as a thickener or
steric stabilizer in aqueous dispersions. In latex suspensions, for
example, the hydrophobic side chains adsorb onto the surface of the
latex particles. Polymers with a high hydrophobe content possess a
large number of sites where these interactions may occur; hence the
free energy required for desorption of a polymer chain may be many
times higher than that for a monomeric surfactant molecule. Once the
side chains have been thus removed from the aqueous suspending
solution, the backbone remains solubilized, keeping the latex particle
suspended as well. Moreover, the presence of the polymer in the
aqueous phase results in a relatively high bulk viscosity.

The interactions between HMHEC and the suspended medium are even
more complex when the suspended medium consists of surfactant
micelles or emulsion droplets. Polymer/surfactant interactions are
important in any system for which both detergency and rheology must
be controlled, as well as in latex suspensions where residual
surfactant is present after the latex-forming polymerization has been
carried out. Examples include shampoo, detergents, cosmetics, other
emulsions, and paints. The complexity of the HMHEC/surfactant

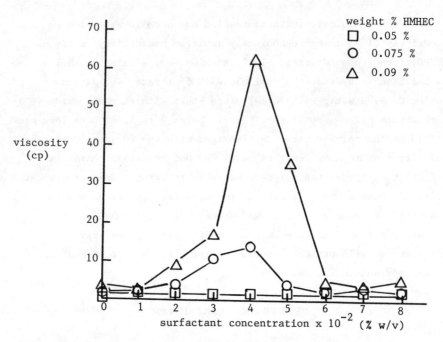

Figure 1. Effect of surfactant concentration on viscosity of
HMHEC / sodium oleate solutions.

interactions arises due to the liquid-like nature of the micelles.
Rather than adsorbing onto the surface of a particulate suspended
phase, the hydrophobic side chains must become incorporated into the
micelles or emulsion droplets themselves. Therefore properties such
as the geometry of the surfactant molecules, their concentration, and
their packing density in micelles or around emulsion droplets will
govern the compatibility of the polymer with a particular surfactant
system.

Figure 1 shows a typical plot of the viscosity of HMHEC/sodium
oleate solutions as a function of surfactant concentration. The
critical micelle concentration (cmc – note: not the same as CMC, the
abbreviation for a water-soluble polymer referred to above) of this
surfactant was determined to be 0.042% (w/v). At low ($\leq$0.05% w/v)
polymer concentrations, HMHEC has no measurable thickening effect on
the surfactant solution, above or below the cmc. At higher polymer
concentrations, the viscosity profiles exhibit a peak at or near the
cmc, and then drop down to a baseline level equal to the viscosity of

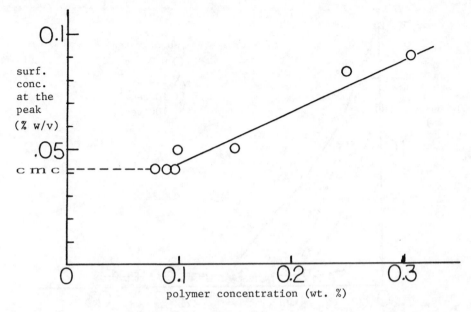

Figure 2. Effect of polymer concentration on the position of the peak in
viscosity profiles of HMHEC/sodium oleate solutions.

the 0.05% polymer solutions. These effects were not observed in
unmodified HEC. As the polymer concentration is increased above
0.095%, the location of the peak moves to higher surfactant levels in
a linear fashion, as shown in Figure 2, with a linear increase
observed in peak viscosity with polymer concentration as well. In
Figure 3 the shear-rate dependence of solutions of HMHEC in Triton X-
100 is shown as a function of the surfactant concentration above the
cmc. The shear-thinning behavior observed at low surfactant
concentrations is characteristic of an entangled network. At higher
surfactant levels no network is in evidence even at the low shear
rates, indicating that the entanglements have been disrupted and the
polymer molecules behave as individual chains.

These results permit us to draw the following picture of
HMHEC/surfactant interactions in these systems. The surfactant
associates with side chains on the polymer via hydrophobic
interactions. The number of units (i.e. surfactant molecules or
individual side chains) involved in each hydrophobic aggregate is
governed by geometric packing constraints which also dictate the size
and shape of pure surfactant micelles. At low surfactant

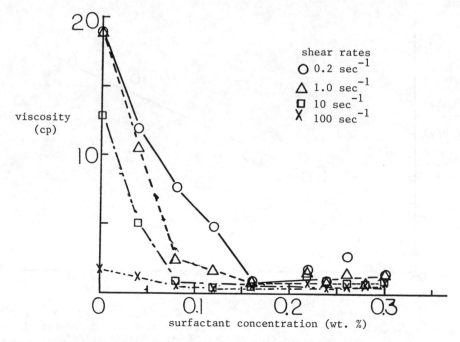

Figure 3.   Shear-rate dependence of HMHEC /Triton X-100 solution viscosity
above the cmc (HMHEC concentration = 0.3 wt. %)

concentrations therefore the aggregates are required to contain
multiple polymer side chains in order to satisfy packing constraints.
This results in the formation of intermolecular "bridges" linking two
or more polymer molecules together and producing a highly viscous
non-Newtonian polymer network. When the surfactant concentration is
increased, the composition of these bridges shifts to a higher
proportion of surfactant molecules and fewer polymer side chains.
The net result is to reduce the extent of entanglement and hence the
viscosity, until a baseline level is reached. If a high
concentration of polymer is present, proportionally more surfactant
is required to solubilize all of the side chains, giving the behavior
shown in Figure 2.

    It is evident from the above discussion that the rheological
properties of HMHEC/surfactant solutions may be varied over a wide
range by adjusting the relative concentrations of the species in
solution. Another performance property which is a direct consequence
of the extent of the network and which is important for applications
such as tertiary oil recovery is the filtration time of the network.

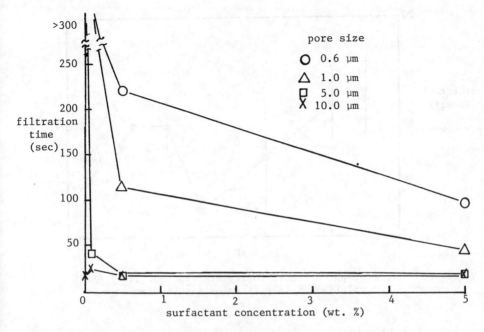

Figure 4. Effect of surfactant concentration on filtration time of
HMHEC /Triton X-100 solutions (HMHEC concentration = 0.7 wt. %)

"Bridged" networks which are stable to shear tend to plug up pores
such as those found in porous rock formations, preventing passage of
the polymer as well as any other material. The extent of plugging
which occurs depends on the composition of the network. Ultimately
this property determines the net flow rate of oil through the pores.

A measure of the shear resistance of the networks was obtained
via filtration experiments of HMHEC/surfactant solutions through
Nuclepore filters of known pore size. Typical results are shown in
Figure 4 for solutions of HMHEC in Triton X-100 above the cmc. The
filtration time through large (>0.8 micron) pores goes down with
increasing surfactant concentration up to 0.5% (w/v) surfactant and
then levels off, indicating that intermolecular aggregation due to
bridging of the side chains has reached a minimum. This is
consistent with the results discussed above. For very small pore
sizes the filtration time is effectively infinite; the limiting pore
size is seen to depend on the surfactant concentration. We conclude
that the extent of the polymer network varies inversely with
surfactant concentration.

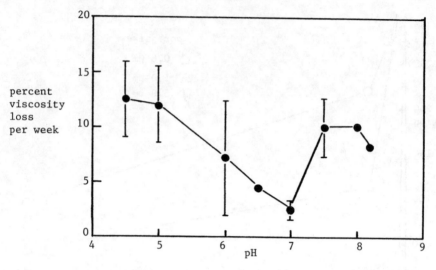

Figure 5. Effect of pH on the viscosity stability of 1.0 % HMHEC solutions in a zwitterionic surfactant.

Studies on the viscosity stability of HMHEC solutions prepared with zwitterionic surfactants [4] have provided evidence that the polymer's compatibility with surfactants is a strong function of the packing density of the surfactant molecules in the micelles. Figure 5 shows the effect of solution pH on the stability of HMHEC solutions in a zwitterionic surfactant with isoelectric range of pH 5.5 - 7. In this range the surfactant head groups are fully charged. In the micelle, like charges are positioned next to each other, leading to maximal repulsion among the head groups and hence a relatively open interface through which hydrophobic side chains from the polymer may enter. The viscosity was measured at weekly intervals during heat-aging of the solutions at 49 ℃ and the slope of the resulting linear function normalized by the initial viscosity to obtain the value plotted on the ordinate. Thus a low relative viscosity loss characterizes a solution in which only thermally stable interactions are present. Clearly the interactions which have the greatest effect on the solution viscosity are those resulting in network formation, while those with the least effect are polymer/micelle interactions involving relatively few polymer side chains. We see that in the isoelectric range very few unstable intermolecular interactions are formed and the viscosity remains relatively constant. Moreover, the initial viscosities in this range average 30-40% less than those outside the range. This means that the tendency for the polymer molecules to form networks to begin with is reduced when the micelles are more open to receive the side chains. Note also that since the stability results are normalized with respect to the initial viscosity, in this region the data shown in the figure are most sensitive to absolute drops in viscosity and still show significant trends toward enhanced stability.

These results may be applied in any system where an entangled network is to be formed in a small or tortuous space. The solution can be prepared initially at the isoelectric pH, poured into the vessel, and then mixed with a small volume of an acid or alkaline solution to modify the structure of the micelles and hence drive some side chains into the bulk phase, where they can aggregate intermolecularly to form a viscoelastic network.

CONCLUSIONS

HMHEC is a versatile cellulosic surfactant which by virtue of its unique structure can interact in a predictable manner with non-aqueous components of dispersions. It can be synthesized with a wide range of side chain levels and lengths and backbone solubilities, making it potentially useful in many applications. Continuing investigations of this polymer are expected to reveal additional insight into the processes of self-aggregation and adsorption in HMHEC solutions.

The authors wish to thank Aqualon Co., Wilmington, De. for permission to publish this work, which was conducted at Hercules, Inc. while C.A.S. was employed there as a member of the technical staff.

REFERENCES

1. Landoll, L.M. J. Polym. Sci., Polymer Chem. Ed. 20:443, 1982
2. Steiner, C.A. Polymer Preprints 26:224 1985
3. Gelman, R.A. and Barth, H.G. Viscosity Studies of Hydrophobically Modified (Hydroxyethyl)cellulose, in ACS Advances in Chemistry Series No. 213, 1986 p. 101
4. Steiner, C.A., manuscript in preparation

DISCUSSION

Q: R. Narayan (Purdue Univ.)

The graft polymer architecture would play an important role. Could you comment on the critical length of the hydrophobe, the degree of substitution, MW of the backbone necessary ?

A: We have seen the same relationship between viscosity and surfactant concentration using C_{12} side chains with C_{12} surfactant and C_{16} side chains with C_8 surfactants. However, the magnitude of the viscosity does increase with increasing side chain length for the same weight % or total moles of side chain. The degree of substitution determines, first, whether or not the polymer is water-soluble, second, the viscosity of polymer/surfactant solutions, and finally, the surfactant concentration at the peak of the function viscosity vs. surfactant concentration. The molecular weight is the major factor influencing the rheological behavior of insoluble HMHECs in surfactant solutions.

STRUCTURE AND APPLICATION PROPERTIES OF POLYELECTROLYTE COMPLEXES

BURKART PHILIPP, HERBERT DAUTZENBERG, HORST DAUTZENBERG,
ULRICH GOHLKE, JOACHIM KÖTZ AND KARL-JOACHIM LINOW
Academy of Sciences of the GDR, Institute of Polymer
Chemistry "Erich Correns" in Teltow-Seehof/GDR

ABSTRACT

Results on formation and structure of polyanion-polycation
complexes (symplexes) in the solid and in the dispersed state
are presented. Symplex application in microencapsulation and
flocculation is considered.

INTRODUCTION

Polyelectrolyte complexes (symplexes) formed by a predominantly Coulombic interaction between polyanion and polycation
to salt-like polymer structures are of scientific relevance
in connection with highly organized polymer structures, and
they are of growing technical importance for processes of
membrane separation, encapsulation, immobilization, coating
and flocculation. An affiliation to cellulose and related polysaccharides is to be seen from the product side in the em-
ployment of ionic cellulose derivatives or ionic polysaccharides as symplex components, and from the process side in the
use of the principle of symplex formation in the cellulose
industry, especially for the treatment of effluents. Starting
from some general considerations on symplex formation and the
routes for varying structure and properties of these systems,
this contribution surveys recent work of our group on symplexes in the solid and in the disperse state together with some
consequences of these results in symplex application, stress-
ing adequately cellulose-relevant problems.

SOME GENERAL REMARKS ON SYMPLEX FORMATION AND STRUCTURE

The main principles of symplex formation, i.e. the reaction
of a free polybase and a free polyacid or between the halide
salt of a polybase and alkali salt of a polyacid are rather
simple and nearly trivial. Nevertheless, these principles exhibit a remarkable versatility and adaptability to special
demands of application, and the variability of symplex structure and symplex properties is very broad. On the other hand,
serious problems can arise in reproducing a definite structure, especially a definite morphology in one and the same
symplex system. This can be traced back to two causes, i.e.
the large number of polyelectrolyte component structures available today and the much higher number of possible combinations on one hand, the multifactual influence of conditions
of symplex preparation on symplex structure on the other.

142

Regarding component structure not only the type of basis polymer and of functional groups, but also charge density,
charge density distribution and molecular geometry (branching) and to some extent also the molar mass of the polymers
have to be considered. Concerning the influence of symplex
preparation, parameters like ionic strength of the system
and polyelectrolyte component concentration must be mentioned in the first place, but also the order of component
addition and the procedure of component mixing can play a
part. The effect of ionic strength (salt content) on symplex
structure and properties will be a major point throughout
this paper. The level of polymer concentration and the degree
of conversion (ratio of cationic to anionic groups) are the
decisive parameters in determining, whether or not a phase
separation occurs, i.e. whether a solid symplex precipitate
(or a symplex film) or a disperse symplex system is finally
obtained. This criterion of phase separation is important
with respect to the applicational properties and will thus
be used subsequently for the systematization of our results.

STRUCTURE AND PROPERTIES OF SYMPLEXES IN THE SOLID STATE

As first shown by Michaels /1/ with styrene-based polyanions
and polycations, stoichiometric polysalts (1:1-symplexes) of
constant composition can be prepared within a wide range of
component ratio given. According to our experience /2/,
"stoichiometric" symplexes with a ratio of cationic to anionic groups between 0.9 and 1.1 are obtainable with many
polycation-polyanion-systems even at rather unequal charge
density of the components at component concentrations of
about 0.1...1 % by weight, if the symplex precipitate is
thoroughly washed.   This is demonstrated in Fig. 1 by the

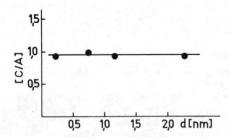

Figure 1. Molar ratio of cationic to anionic groups (C/A) in
symplexes between carboxymethyl cellulose (DS ≈ 1) and cationically modified polyacrylamides with different distance
d between cationic charge centres, prepared with a large excess of the cationic component

composition of symplexes between carboxymethyl cellulose and
cationically modified polyacrylamides of different charge
density. Significant deviations from this 1:1 stoichiometry

were observed in the case of extreme misfit between the components for charge compensation or in the case of inacti-
vation of one of the ionic groups by protonation or deproto-
nation.

As revealed by X-ray and NMR-investigations of ours and
of other groups, the supermolecular structure of isolated solid symplexes can be considered to lie in between the two
borderline models of a chaotic entanglement of anionic and
cationic chains with a statistical charge compensation
("scrambled-egg model" according to Michaels) and a well-ordered ladder structure with defined Coulombic bonds (comp.
Fig. 2). The submicroscopic morphology of symplexes in the

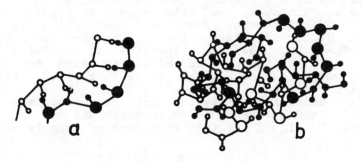

Figure 2. Models of symplex structure   a. ladder model
                         b. scrambled-egg model

solid state varies within very wide limits in dependence on
polyelectrolyte system employed and the conditions of symplex
precipitation and can hardly be exactly controlled up to now,
while the gross morphology, of course, is determined by the
external parameters of preparation.

The colorless, horny or brittle film-forming products
obtained, are stable at temperatures of about 100 °C and nearly reversibly swellable in water, maintaining considerable
mechanical strength also in the swollen state. In the case
of toxic components being applied, this toxicity is generally
significantly diminished or even eliminated by the symplex
formation.

APPLICATION OF SYMPLEXES IN THE SOLID STATE, ESPECIALLY
                    FOR MICROENCAPSULATION

Due to their structural versatility, symplex products have
been proposed for application in a large number of areas,
especially in the biomedical field. In connection with the
high hydrophilicity and swellability, good mechanical
strength in aqueous media and low toxicity, an application
in membrane and membrane processes, especially for use in
the biomedical field, seems to be very promising and has
occasionally already been realized commercially. Besides the
preparation and/or surface modification of flat membranes or

hollow fibres, the application of symplexes in microencapsulation of biological substrates is a challenging scientific
and technical problem. Two routes are known here today, i.e.
a multi-step process employing alginate and polylysin as symplex components (3), and a one-step process developed in our
institute by Dautzenberg et al. (4); which will now be considered in some detail. The general procedure of microencapsulation consists of a fast interfacial reaction between a
drop of an aqueous solution of the anionic polymer containing
the substrate and an aqueous solution of the cationic polymer
acting as a "precipitation bath". Although capsules can be
obtained from several different polyanion-polycation combinations, only a system composed of low DS, completely water-
soluble Na-cellulose sulphate and of poly(dimethyldiallyl-
ammonium)chloride ("PDMDAAC") was found so far to be really
suitable for practical use, with respect to capsule wall morphology - in consequence of this - of appropriate mechanical
strength and permeability to low and high molecular compounds. These applicational properties are controlled in a
rather sensitive manner by the molecular parameters of the
polyelectrolyte components and the conditions of reaction
like polymer concentration and time of residence in the precipitation bath. A decisive influence on capsule wall properties is exerted by the ionic strength in the system, as demonstrated in Fig. 3 for the dependence of capsule strength

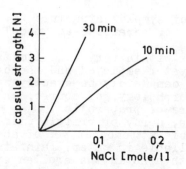

Figure 3. Mechanical strength of symplex microcapsules in dependence of NaCl-content in the component solutions at 10 and
30 min residence time in the precipitation bath

on the NaCl-content in the cellulose sulphate solution and
the precipitation bath at two different residence times. In
the range of salt concentration considered here a rather
steep increase in strength with the amount of NaCl added can
be concluded from this Figure. This change in mechanical
strength can be correlated with significant structural
changes on the submicroscopic level. The diameter of the sphe-
rical capsules can be varied between about 200 and 5000 µm;
the capsule wall shows an asymmetric morphology and thickness
between about 2 and 20 µm (comp. (5)). Special advantages of
this route to microencapsulation are the "quasi-physiologi-

cal" conditions of reaction and the compatibility of special
grades of cellulose sulphate and of the symplex forming the
c$_a$psule wall with many sensitive biological materials. Table
1 gives some examples of compounds and systems that could be
successfully encapsulated without significant loss of biological activity.

TABLE 1
Application of symplex microencapsulation

Substrate	Area of application
enzymes (urease)	blood detoxication
cell fragments (liver micro-somes)	blood detoxication
cells (ovarial cells)	cattle breeding
tissue (pancreas)	treatment of diabetes

SOME STRUCTURAL FEATURES OF DISPERSE SYMPLEX SYSTEM

Soluble or quasisoluble symplexes without macroscopic phase
separation can be prepared along two routes, i.e. as "sequential symplexes" by covering very long polyions (I) partially
with shorter ones (II) of opposite charge via Coulombic bonds
with the ratio of $M_I:M_{II}$ being about 10:1, and by reacting
oppositely charged polyions of about equal length in highly
dilute solution ($\leq$ 0.1 % by weight) at low or medium degree
of conversion. The fast route started by the pioneering work
of Tsuchida and Kabanov is leading now to application in the
field of immunology. The second approach persued by us is relevant with regard to flocculation.

Our combined light scattering, sedimentation and viscosity measurements with various polyanion-polycation systems
led to the conclusion (2) that generally not single "symplex
molecules", but symplex aggregates are formed, the size and
coil density of which depends on the type of ionic groups, on
the level of charge density and on the charge density difference between the components (comp. Table 2). In the salt-free
systems, the size of the non-stoichiometric, quasi-soluble
symplex aggregate remains nearly constant between 20 and
60...70 % conversion, and only the number of aggregates increases.

On adding NaCl to these systems, 3 types of response
with increasing ionic strength can be discerned, i.e.
(i) a further aggregation and subsequent precipitation due
   to a "salting-out" effect in systems like PDMDAAC/cellu-
   lose sulphate or PDMDAAC/lignosulphonate.
(ii) rather small changes in system structure only due to a
   compensation of the salting-out effect and a counter-
   acting shielding effect
(iii) a gradual swelling and dissolution of the symplex particles, until a complete disaggregation occurs at a definite, charge density dependent "critical" salt concentration due to a predominant polyelectrolyte shielding effect, as realized with systems composed for example of anionically and cationically modified polyacrylamides.

TABLE 2
Aggregate size and coil density of quasisoluble symplexes

System	$c_{symplex}$ g/cm^3.10^4	macromolecules particle	coil density g/cm^3.10^4
PDMDAAC/cellulose sulphate	0.35	2000	165
PDMDAAC/polyacrylate	0.69	1860	421
PDMDAAC/anionic poly- acrylamide of low charge density	2.12	8	2
cationic polyacrylic ester/polyacrylate	1.86	200	91
cationic polyacrylamide/ anionic polyacrylamide, both of low charge density	4.22	15	14.9

Also with respect to the actual problem of "preferential binding" in systems containing more than two polyelectrolytes the ionic strength can play a major role. An example is given by the system lignosulphonate, carboxymethyl cellulose/ PDMDAAC. In a salt-free system, CMC is preferentially bound to PDMDAAC at low degree of conversion, and at medium degree of conversion both polyanions are incorporated into the symplex, while in presence of NaCl a strong preference for lignosul- phonate is observed.

At a somewhat higher polymer concentration of about 0.1...0.5 % by weight the transition range from the quasiso- luble symplex to complete phase separation can be rather con- veniently investigated in dependence on degree of conversion by a combination of turbidimetric, conductometric and poten- tiometric titration techniques (2). The titration endpoints usually indicate a non-stoichiometric symplex composition de- pending on kind of functional groups, charge density and mo- lecular geometry of the polyions. This non-stoichiometry is obviously at least partially caused by surface charges on the particles, i.e. an uneven charge distribution in the symplex, as shown by our electrophoretic measurements with systems of anionically and cationically modified polyacrylamides (6). Conductometric and turbidimetric endpoints usually coincide in systems of two linear components, while deviations are obser- ved, if one of the polyions is branched. Particular colloidal phenomena were found in symplex formation with regular branched polyanions consisting of a backbone of variable hy- drophobicity and rather long alkyl sidechains with a carboxy- lic group at their end (7). In some systems of this kind spon- taneous formation of a homogeneous gel was observed, while with others a stable symplex dispersion was obtained, indi- cating that besides kind of ionic groups and charge density also the hydrophil/hydrophob ratio and the molecular geometry of the polymers play a part in determining the state of dis- persity. Adding titrant to the system beyond the titration end-point, further and rather complex changes in the state of dispersion (flocculation, redissolution or coacervation) were noticed in some of these and also in other systems, which may

be connected with changes in the number and even the sign of
surface charges.

Special features are encountered in symplex formation
with polyampholytes. As shown with copolymers of maleic acid
and allylamine derivatives of different N-functionality,
"homo-symplex" formation between acid and basic groups can
lead, but must not necessarily lead, to phase separation near
the isoelectric point, depending on the kind of N-function.
Heterosymplex formation was feasible here with cationic polyelectrolytes only, with some competition between homo- and
heterosymplex formation obviously occuring with strong cationic polyelectrolytes like PDMDAAC.

The effect of added salt on disperse symplex systems in
the transition range to phase separation is similar to that
observed with quasisoluble symplexes. In many systems of linear polyelectrolytes of low to medium charge density containing one "weak" polyelectrolyte the turbidity decreases
with increasing ionic strength during the whole course of titration, and above a "critical" salt concentration a clear
one-phase system is maintained without visible symplex aggregates being formed (comp. Fig. 4). In some special systems

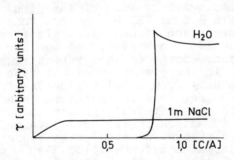

Figure 4. Turbidity in dependence of the molar ratio of cationic to anionic groups (C/A) of a branched polycarboxylate
anion with cationic polyacrylic ester in water and in 1 m
NaCl

like polystyrene-sulphonate/cationic polyacrylamide with quaternary ammonium groups, a shielding effect of the salt impe-
des precipitation after reaching a turbidity maximum during
the titration and leads to a stable dispersion.

On the other hand, flocculation of a disperse symplex
system can be promoted by addition of salt, if hydrophobic
forces contribute significantly to interaction between the
components. Generally, the effects of added salt on disperse
symplex systems in this range of higher polymer concentration
are of a similar complexity as in the case of quasisoluble
symplexes.

APPLICATION OF SYMPLEX FORMATION TO FLOCCULATION

The principle of symplex interactions can be applied by two
ways to achieve or promote the separation of suspended par-

ticles and for some categories of dissolved compounds from
aqueous systems, especially from effluents, i.e. either by
direct symplex formation between the compound to be eliminated and a suitable polyelectrolyte added, or by adding appropriate symplex precursors, which on subsequent symplex formation in situ act as a primary flocculant by incorporating
the colloidal particles into and between the symplex aggregation. A combination of both principles is feasible. Besides
Coulombic forces, other types of interaction like hydrophobic or polar binding often is effective in substrate-floccu-
lant interaction. These general statements shall be illustrated now by some examples from our recent work.

A concentration of lignosulphonate from dilute aqueous
solution in pulp mill effluents can be achieved by binding
the anionic polyelectrolyte to linear poly(ethyleneimine) to
form a symplex precipitate at pH $\leq$ 7, which after separation
is decomposed in an alkaline medium to the free PEI base as
a crystalline phase and a concentrated lignosulphonate solution. Suitable symplex precursors can be realized either
by mixing adequate polyelectrolytes at external conditions
impeding symplex precipitation, or by employing appropriate
polyampholytes. In both cases flocculation occurs by symplex
formation in situ. With systems of the first kind employing
for example again linear PEI as the cationic component, an
efficient sedimentation of colloidal dispersed kaolin, $CaCO_3$
or pulp fines could be achieved, and by providing an anionic
or cationic excess charge on the symplex aggregates via the
component ratio in the precursor, ionic dyes could be separated simultaneously from model effluents. The amount of a
symplex precursor required is in the range of 0.1 - 10 ppm.

A versatile and efficient primary flocculant operating
by homosymplex formation in situ was obtained by us via an
aminolysis of poly(acrylonitrile) with dicyandiamide (8).
The resulting polyampholyte (Awifloc 311) contains in a molar basis 62.5 % neutral amide, 15.6 % acidic, 11.6 % weakly
basic and 4 % strongly basic groups. It is applied as a
clear, weakly acid aqueous solution forming homosymplex pre-
cipitates from pH 3 to pH 9, which also has a considerable
binding capacity for several categories of dissolved compounds via polar or hydrophobic forces. As an example the
separation of chlorinated lignin fragments from sulphite
pulp bleaching waste water is given in Table 3. The Awifloc
system was successfully employed to achieve separation
in effluent compositions where conventional flocculant
failed, and it can be further modified to meet special demands.

TABLE 3

Separation of chlorinated lignin fragments from aqueous solution by Awifloc(R)

Amount of Awifloc	% Lignin elimination
100 ppm Awifloc 311	31
200 ppm Awifloc 311	73
400 ppm Awifloc 311	77
800 ppm Awifloc 311	79
100 ppm Awifloc + 2 ppm PDMDAAC	64

REFERENCES

1. Michaels, A.S. and Miekka, R.G., Polycation-polyanion complexes: Preparation and properties of poly(vinylbenzyltri-
   methylammonium)poly(styrenesulphonate). J. Phys. Chem.,
   1961, 65, 1765-73.

2. Philipp, B., Dautzenberg, He., Linow, K.-J., Kötz, J. and
   Dawydoff, W., Polyelectrolyte complexes - recent development and open problems. Progr. Polym. Sci. (in press).

3. Lim, F., Verfahren zum Einkapseln von chemischen oder bio-
   logisch aktiven Stoffen, lebendem Gewebe und Einzelzel-
   len. DE-OS 30 12 233.

4. Dautzenberg, Ho., Loth, F., Pommerening, K., Linow, K.-J.
   and Bartsch, D., Mikrokapseln und Verfahren zu ihrer Herstellung. DE-OS 33 06 259.

5. Dautzenberg, H., Loth, F., Wagenknecht, W., and Philipp,
   B., Cellulose - Ausgangsmaterial für hochveredelte Pro-
   dukte im biologisch-medizinischen Bereich. Das Papier,
   1985, 39, 601-07.

6. Kötz, J., Philipp, B., Sigitov, V., Kudaibergenov, S. and
   Bekturov, E.A., Amphoteric character of polyelectrolyte
   complex particles as revealed by isotachophoresis and vis-
   cometry. Colloid and Polymer Sci. (in press)

7. Kötz, J., Linow, K.-J., Philipp, B., Vogl, O. and Hu,
   L.P., Some effects of charge density and branching on the
   composition of polyanion-polycation-complexes. Polymer
   (London), 1986, 27, 1574-80.

8. Gohlke, U., Dietrich, K., Bischoff, C., Versäumer, H.,
   Hartig, S., Wolf, K.-F., Liebener, U. and Rudolph, J.,
   Acrylnitrilcopolymere und Verfahren zu ihrer Herstellung.
   EP 01 48 295.

DISCUSSION

Q: H. Inagaki (Kyoto Univ.)
Have you tried to make polyion complexes using anionic
and cationic cellulose derivatives ?

A: So far we only combined anionic cellulose derivatives,
i.e. carboxymethyl cellulose, carboxyethyl cellulose and
cellulose sulphate in a rather wide range of DS with synthetic cationic polyelectrolytes. For investigation on polyelectrolyte complexes with the same backbone chain in the anionic
and the cationic component we employed anionically and cationically modified polyacrylamides of different charge density.

Q: J. Komiyama (Tokyo Inst. Tech.)
Can you form pores in your films or microcapsule surfaces ? If so can you control the size ?

A: Generally, the pore structure of cellulose based films
and membranes can be influenced by the molecular parameters
of the polymer, the procedure of film formation and the conditions of after treatment.
Concerning symplex microencapsulation, permeability and
mechanical strength depending on capsule wall morphology can
be controlled in a qualitative way via the molecular parameters of the symplex components, the polymer concentration
of substrate solution and precipitation bath, the concentration of low molecular electrolytes like NaCl, and the time of
residence of the capsules in the precipitation bath.

LANGMUIR-BLODGETT FILMS OF CELLULOSE DERIVATIVES

MUTSUO MATSUMOTO*, TAKAHIRO ITOH** and TAKEAKI MIYAMOTO*
* Institute for Chemical Research, Kyoto University
Uji, Kyoto-Fu 611, Japan
** Tokai Senko K.K., Mibutsuji-Machi, Nakagyo-Ku,
Kyoto 604, Japan

ABSTRACT

Monolayers of cellulose tridecanoate(CTD) and cellulose trioctadecanoate(CTO) were prepared on the water surface. The monolayer of CTD exhibited an apparent plateau region in the surface pressure(π)-area(A) isotherm, whereas no plateau was observed in the π-A isotherm of CTO monolayer. This suggests that the formation behavior of cellulose derivative monolayers depends on the length of alkyl groups in the side chains. The fine structure of the films transferred at sequential stages of surface pressures was directly examined by using the dark-field and bright-field imaging modes of electron microscope. It was found for CTD that the hetero-geneous film, which had included a number of large holes at early stages of surface pressures, was converted to a homogeneous two-dimensional film on compression and that further compression led to the formation of bilayer film and eventually to a collapsed film. On the other hand, CTO showed that on compression, the island film containing fine holes with irregular shapes was converted to a two-dimensional homogeneous film and then led directly to a collapsed film on further compression. The results were discussed in terms of the difference in intermolecular dispersion forces of the side chains.

INTRODUCTION

From the practical and academic points of view, interest has steadily developed in the Langmuir-Blodgett(LB) films of polymers. The advantage of the LB method is that it is possible to make well-ordered uniform films controllable to a molecular dimension. However, it is known that monolayers of low-molecular-weight amphi-philic compounds such as stearic acid are not always homogeneous(1). In order to study the structure-property relationship of LB films, it is important to elucidate the fine structure of monolayers. Among various methods employed to analyze the monolayer structure such as spectroscopic methods, surface potential measurement, X-ray and electron diffraction and so on, the electron microscopy will be most suitable to the direct observation of the fine structure of film surfaces. The fine structure of monolayer can be observed by using the dark-field imaging mode of electron microscopy, as successfully applied to structures of stearic acid monolayers(1), while, the thickness of the monolayers can be examined by the bright-field electron microscopy(2,3).

Cellulose is a hydrophilic, linear, stereoregular macromolecule and can be easily converted into amphiphilic derivatives by introducing hydrophobic substituents such as alkyl groups. Cellulose is of practical interest as a material suitable for preparing functional LB films. With regard to mono- and multi-layers of cellulose derivatives, however, only a few studies have been reported on the π-A characteristics(4). No detailed study has been made on the structural characteristics, especially, the fine structures of cellulose derivative monolayers.

Recently we have been engaged in the studies on the relationships between the structures of both hydrophilic and hydrophobic units to the properties of the cellulose derivative LB films. This report will be concerned only with our studies on the fine structures of monolayers from cellulose derivatives. The monolayers of cellulose tridecanoate(CTD) and cellulose trioctadecanoate (CTO) were spread on the water surface and the fine structures of the films transferred at sequential stages of surface pressures were directly examined by using the dark-field and bright-field imaging modes of electron microscope.

EXPERIMENTAL

Cellulose Derivatives
Cellulose triester samples CTD and CTO were prepared from regenerated cellulose by the acid chloride-pyridine procedure. The regenerated cellulose was prepared from cellulose acetate having a viscosity-average degree of polymerization of ca. 200 by treatment with aqueous ammonia. The details of the experimental procedure have been described elsewhere(5). The degree of substitution(DS) of the sample derivatives was determined according to the conventional back-titration method and found to be 3.0 for both samples. The DS values were also estimated with ^{1}H-NMR spectroscopy, where the results were consistent with those by the back-titration method within an experimental error.

Preparation of Monolayers
The trough of rectangular type(200 x 500 x 3 mm), made of Teflon-coated aluminum, was kept at 293 K by circulating thermostated water. The preparation of monolayers followed a common method(6); after sweeping a water surface carefully with a Teflon-coated barrier, the sample derivative was spread on a clean water surface from the benzene solution and stood for about 30 min. till complete evaporation of the solvent. Solutions for spreading monolayers were prepared by dissolving a few miligrams of the sample in 5 cm^3 of benzene (analytical grade). The monolayer was compressed with a Teflon-coated barrier at a speed of 4 mm· min^{-1}. The surface pressure was recorded by the Wilhelmy method.

Dark-Field and Bright-Field Images of Transferred Films
The specimens for the observation of surface structures were prepared by lifting the surface film horizontally to the specimen grids covered with hydrophobic carbon supporting film. Namely, the carbon film was positioned in parallel to the mono-layer surface, touched to the monolayer surface from air side and then withdrawn using a motor driving apparatus. By this operation, the air side, i.e., the hydro-phobic portion of monolayer adheres directly to the hydrophobic carbon supporting

film. The water drop attaching to the monolayer was removed with a filter paper. The surface structure of the films transferred on the carbon supporting film was observed by applying the dark-field imaging mode of a JEM 200-CX electron microscope at the direct magnification of 2,800 times. The minimum dose system was used to reduce the radiation damage to the sample films.

For the film thickness determination, a specimen grid covered with hydrophilic platinum supporting film was used. Here the specimen grid was suspended vertically in water of the trough in prior to spreading the monolayer. After spreading and compressing the monolayer, the specimen grid was drawn up through the monolayer surface at a speed of 1 mm·min^{-1}. This procedure allows the single layer film of the monolayer transferred onto the platinum supporting film where the hydrophilic portion of monolayer adheres. The thickness of the films transferred was recorded by using the folding method of electron microscope. The details of the procedure have been described elsewhere(2). The film lined with a platinum supporting film was folded into two with the lining inside. The folded edge was observed by a JEM 200-CX electron microscope. The thickness of the sample films was directly estimated in bright-field imaging mode at the direct magnification 160,000 times, because of the clear difference in diffraction contrast between the cellulose and platinum films.

RESULTS AND DISCUSSION

Surface Pressure-Area Isotherm

Figure 1 shows the π-A isotherms of the samples CTD and CTO. It can be seen that the monolayer of CTD derivative exhibits an apparent plateau region in the π-A isotherm, whereas there appears no plateau region for CTO derivative. The isotherms were reproducible. Here, it should be noted that there exists a small maximum in the plateau region of CTD monolayer (point c). The surface pressure began to increase when the monolayer was compressed to a point where the area is about 0.9 nm^2 per glucose unit (point a). Similar experiments have been carried out by Kawaguchi et al(4). The surface pressure corresponding to the point a and the length of the plateau were different from those obtained here. Furthermore they observed no such maximum in the π-A isotherm of CTD. The reason for this discrepancy is not clear at present, but probably due to the difference in the temperature(293 K for the present work, 290 K for Kawaguchi et al.) at which the measurements were carried out and/or the difference in the solvent used for spreading(benzene for the present work, chloroform for Kawaguchi et al.)(4). On the other hand, the π-A isotherm of CTO monolayer began to rise at an area of about 0.9 nm^2 per glucose unit and then followed a rapid increase in the surface pressure. In the case of CTO, the curve obtained was nearly the same as that reported by Kawaguchi et al.(4).

In general, the π-A isotherms of polymer monolayers spread at the air-water interface are rather featureless when compared with the curves which have been obtained for monolayers of small molecules(6). However, the incorporation of long hydrocarbon side chains on the polymer chain is expected to enhance the film coherence, and the compressibility of the films may approach to that of condensed monolayer films of monomeric paraffin-chain compounds. An apparent plateau appears in the π-A isotherms of monolayers of synthetic polypeptide such as poly(γ-methyl-L-glutamate) and poly(γ-benzyl-L-glutamate)(2,3,7). Here this plateau region is considered to reflect a transition process from monolayer to bilayer film (2,3,7). It is of interest to note that the monolayer of CTD having a relatively

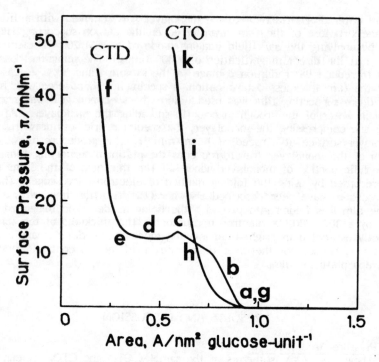

Figure 1. Surface pressure(π)-area(A) isotherms of cellulose tridecanoate(CTD) and cellulose trioctadecanoate(CTO) monolayer, spread from benzene solution at 293K.

short alkyl substituent exhibited a similar π-A isotherm to those of synthetic polypeptide monolayers, whereas the monolayer of CTO having long alkyl substituents to those of long-chain fatty acids such as stearic acid. It is obvious that the difference in monolayer formation between these cellulose derivatives is due to the difference in intermolecular dispersion forces among side chains.

Fine Structures of the Monolayers

The dark-field imaging mode is performed by shifting the objective aperture to block the main electron beam and, at the same time, to admit the electrons scattered from the sample. Therefore, the sample monolayer appears as bright images on a dark background, while the situation is reverse in the case of the bright-field imaging mode. Figure 2a shows a typical feature of CTD monolayer which appeared at the point a where the area A= 0.95 nm^2 per glucose unit and π = 0 mN·m^{-1}, together with an electron micrograph at the folded edge of the monolayer film transferred to the platinum supporting film. A number of irregular holes ranging from 200 to 900 nm in diameter are seen in the surface film. The film could not be observed in a bright-field imaging mode, indicating that the film was not within a measurable thickness. Figure 2b shows the results at the point b where the pressure π = 6 mN·m^{-1} and the area A= 0.83 nm^2. The size of vacant areas decreased and the thickness of the film was found to be ca. 0.5 nm, which is approximately equal to the thickness of glucose ring lying flat on the water surface. Figure 2c shows the electron micrographs at the point c where the pressure $\pi = 14$ mN·m^{-1} and the area A= 0.59 nm^2. It can be seen that a completely homogeous monolayer film has been obtained at point c. The thickness of the film is in fairly good agreement with that of the molecular model of CTD, in which the glucose ring lies flat on the water surface and the alkyl groups in the side chains are directed normal to the glucose ring(Figure 3). Figure 2d shows an electron micrograph of the surface film in the plateau region. The film contains a number of star-like spots uniformly scattered all over the film, showing that the film became again heterogeneous. The electron micrographs of the film at point e, where the second steep rise was observed in the π-A isotherm, are shown in Figure 2e. The thickness of the film was found to be 2.5 nm. This means that the film is approximately twice as thick as the monolayer film at point c. Taking into account the fact that the area at point e was just half a value of monolayer film at point c, the results confirmed that the bilayer was formed at point e. This is consistent with the view that the plateau region in the π-A isotherm of polypeptide monolayers reflects a transition process from monolayer to bilayer(2,3,7), the molecular conformation as well as orientation being left unchanged. However,

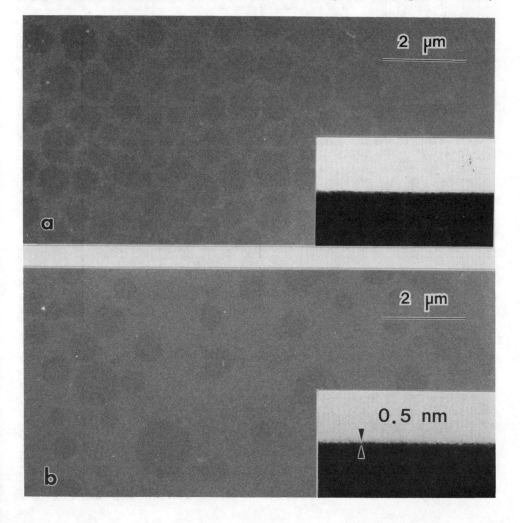

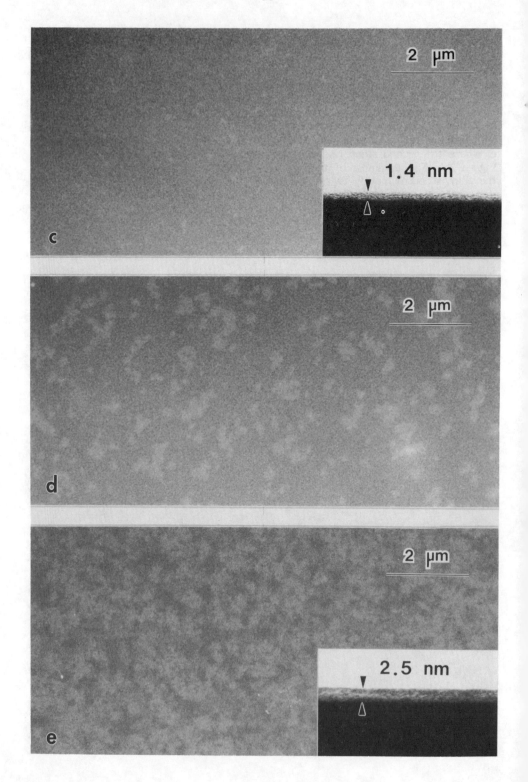

Figure 2. Dark and bright(inset) field images of cellulose tridecanoate monolayers at sequential stages of π-A isotherm ; a, $\pi = 0$ mN·m^{-1} and A= 0.95 nm^2 ; b, $\pi = 6$ mN·m^{-1} and A= 0.83 nm^2 ; c, $\pi = 14$ mN·m^{-1} and A= 0.59 nm^2 ; d, $\pi = 13$ mN·m^{-1} and A= 0.46 nm^2 ; e, $\pi = 15$ mN·m^{-1} and A= 0.30 nm^2 ; f, $\pi = 42$ mN·m^{-1} and A= 0.19 nm^2.

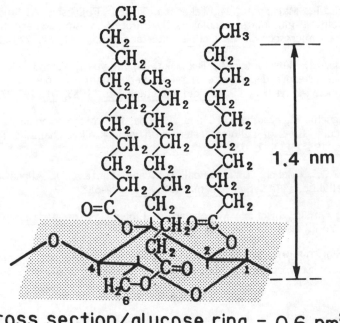

cross section/glucose ring = 0.6 nm^2

Figure 3. A molecular model for cellulose tridecanoate monolayer.

the mechanism of bilayer formation and its structure are not clear at present. Figure 2f shows the result at point f. This figure indicates that the film collapse on further compression.

As already mentioned, the behavior of π-A isotherm of CTO monolayer was simple. The fine structure and thickness of the monolayer were examined at four points indicated in the π-A isotherm. A heterogeneous film composed of large aggregates in different sizes(ca. 1 to 10 μm in diameter) was found to be formed at point g where the surface pressure $\pi = 0$ mN$\cdot$m^{-1} and area A= 0.94 nm^2. A homogeneous monolayer film was observed at point h where $\pi = 10$ mN$\cdot$m^{-1} and A= 0.70 nm^2. The thickness of the monolayer film was ca. 2.4 nm, being in good agreement with the value of 2.6 nm, i.e., the thickness estimated from the same molecular model as that for CTD. Observations at point i and k revealed that the partial collapse of homogeneous monolayer film occurred. In conclusion, it may be said that homogeneous monolayer films can be obtained from cellulose derivatives by controlling the chemical structure of substituents and the preparation conditions of the films. The difference in monolayer formation between cellulose derivatives may be ascribed to the difference in cohesive force between the alkyl substituents.

REFERENCES

1. Uyeda, N., Takenaka, N., Aoyama, K., Matsumoto, M. and Fujiyoshi, Y., Holes in a stearic acid monolayer observed by dark-field electron microscopy. Nature, 1987, 327, 319-321.

2. Takeda, F., Matsumoto, M., Takenaka, T. and Fujiyoshi, Y., Studies of poly-γ-methyl-L-glutamate monolayers by infrared transmission spectroscopy and electron microscopy. J. Colloid Interface Sci., 1981, 84, 220-227.

3. Takeda, F., Matsumoto, M., Takenaka, T., Fujiyoshi, Y. and Uyeda, N., Surface pressure dependence of monolayer structure of poly-γ-benzyl-oxycabonyl-L-lysine. J. Colloid Interface Sci., 1983, 91, 267-271.

4. Kawaguchi, T., Nakahara, H. and Fukuda, K., Monomolecular and multi-molecular films of cellulose esters with various alkyl chains. Thin Solid Films, 1985, 133, 29-38.

5. Malm, C.J., Mench, J.W., Kendall D.L. and Hiatt, G.D., Aliphatic acid esters of cellulose. Ind. Eng. Chem., 1951, 43, 684-688.

6. Gains, G.L.Jr., Insoluble Monolayers at Liquid-Gas Interfaces, Interscience, New York, 1966.

7. Malcolm, B.R., Molecular structure and deuterium exchange in monolayers synthetic polypeptides. Proc. Roy. Soc. London A, 1968, 305, 363-385.

DISCUSSION

Q: J. Hayashi (Hokkaido Univ.)
Are there any orientation of cellulose chain ?

A: The concrete discussion on the orientation of cellulose chains (main chains) is not available from the results obtained here.

Q: H. Inagaki (Kyoto Univ.)
You found a nice experimental evidence for a model that the side-chain molecules are standing perpendicular to the membrane surface. Can you explain the reason for the model from thermodynamic standpoint ?

A: The surface pressure reflected the intermolecular force depends on the density of monolayer molecule. Therefore, to give a high surface pressure like the point C shown in Fig.2-C, the side chain should orient normal to the water surface, when the van der Waals dispersion component in surface pressure is discussed.

Q: K. Shimamura (Okayama Univ.)
Have you estimated difference in the contrast between the substrate and your cellulose derivatives ? From my experience, your contrast difference looks too much.

A: To record the film thickness of LB film by the bright field imaging mode of electron microscope, the monolayer of cellulose derivatives was transferred on the platinum supporting film due to the dipping out procedure. The LB film transferred on the supporting film was folded into two with the lining inside and then the image of the LB film was recorded at the folded edge by the electron microscope. The clear difference in image contrast was thus obtained between the cellulose derivative LB and platinum films.

Q: T. Komoto (Gunma Univ.)
(1) Did you use a cold or cryo-stage for TEM observation?
(2) Did you find out something change in molecular ordering by annealing of the LB film ?
(3) Did you carry out the staining method to get a finer morphology of the LB film ?

A: (1) We applied the cryoelectron microscopy to confirm the holey structure of fatty acid LB film, for no sublimation of LB film was present. The image taken by cryoelectron microscope was entirely the same as image taken by dark field imaging mode of electron microscope.
(2) We never tried the annealing on cellulose derivative LB films. This treatment will be expected to change the fine structure of LB film.
(3) The negative staining is applicable to find the structure of micelles, vesicles and LB films. However, heavy metals in the staining solution may affect to the change of

structure of LB film by ion binding with ionized polar groups during the staining process. Hence, the application of this method to LB film should be carefully used.

Summary: M. Matsumoto (Lecturer)

The surface pressure reflected the intermolecular force between water and monolayer molecules is directly connected to the molecular orientation and structure of monolayer. For the practical applications such as chemical sensors and mole-cular electro device the monolayer should be transferred on the solid substrate at the well-defined state of monolayer, that is at the two-dimensionally uniform state. We believe that the dark field imaging mode of electron microscope is most useful method to directly inspect the fine structure of LB film. The single layer LB film prepared by cellulose ester tridecanoate forms two-dimensionally uniform film with the orientation of side chains normal to the surface, and hence the further substitution of functional groups to cellulose derivatives will give the possibility as an advanced material as far as the same uniformity of LB film is maintained.

OPTICAL RESOLUTION BY POLYSACCHARIDE DERIVATIVES

YOSHIO OKAMOTO, RYO ABURATANI, KAZUHIRO HATANO,
YURIKO KAIDA AND KOICHI HATADA
Department of Chemistry, Faculty of Engineering Science,
Osaka University, Toyonaka, Osaka 560, Japan

ABSTRACT

Substituted phenylcarbamate derivatives of cellulose, amylose, chitosan, xylan and dextran were prepared and used as chiral stationary phases for high-performance liquid chromatography to separate racemic compounds. Optical resolving power of the phenylcarbamate derivatives depended very much on the substituents introduced on the phenyl groups. Cellulose or amylose derivatives usually exhibited the highest optical resolving ability to most racemic compounds, and their tris(3,5-dimethylphenylcarbamate)s gave practically useful chiral stationary phases.

INTRODUCTION

Recently, optical resolution by high-performance liquid chromatography (HPLC) has been attracting much attention as a practically useful method for obtaining optical isomers and determining their purity, and many chiral stationary phases (CSP) have been reported in the past ten years [1]. Optically active polymers have been used as CSPs. These include naturally occurring macromolecules such as proteins and cellulose, and synthetic polymers such as polymethacrylates and polyacrylamides. We have been studying the CSPs consisting of phenylcarbamate derivatives of cellulose and other polysaccharides [2-4]. In this article, the optical resolution with cellulose trisphenylcarbamate derivatives (1-17) and the 3,5-dimethyl-phenylcarbamate derivatives of amylose (18), chitosan (19), xylan (20) and dextran (21) is mainly described.

$$X = \begin{array}{lll} 1:4\text{-}CH_3O & 7:4\text{-}F & 13:2\text{-}CH_3 \\ 2:4\text{-}CH_3 & 8:4\text{-}Cl & 14:3,4\text{-}(CH_3)_2 \\ 3:4\text{-}CH_3CH_2 & 9:4\text{-}Br & 15:3,5\text{-}(CH_3)_2 \\ 4:4\text{-}(CH_3)_2CH & 10:4\text{-}CF_3 & 16:2.6\text{-}(CH_3)_2 \\ 5:4\text{-}(CH_3)_3C & 11:4\text{-}NO_2 & 17:3,4,5\text{-}(CH_3)_3 \\ 6:\ \ H & 12:3\text{-}CH_3 & \end{array}$$

Amylose	Chitosan	Xylan	Dextran
18	19	20	21

MATERIALS AND METHODS

The polysaccharide phenylcarbamate derivatives, 1-21, were prepared by the reaction of polysaccharides with corresponding phenylisocyanate derivatives. ^{1}H-NMR and elemental analyses indicated that the hydroxy groups were almost quantitatively converted to urethane groups. The derivatives were adsorbed on macroporous silica gel (particle size 10 μm, pore size 400 nm) which had been treated with 3-aminopropyltriethoxysilane; the weight ratio of the carbamate to silica gel was 25:100. Each of the packing materials obtained was packed in a stainless steel tube (25 x 0.46 (id)cm) by a slurry method. Chromatographic analysis was done on a JASCO TRIROTAR-II equipped UV and polarimetric detectors using a hexane-2-propanol (90:10) mixture as an eluent at a flow rate of 0.5 ml/min at 25 °C. Dead time (t_0) was determined with 1,3,5-tri-tert-butylbenzene.

RESULTS AND DISCUSSION

Figure 1 shows the chromatogram of the optical resolution of Tröger base (22) on a cellulose tris(4-tert-butylphenylcarbamate) (5) column. Complete optical resolution is attained and (-)-isomer elutes at t_1 and (+)-isomer at t_2. Then, capacity factors, $k'_1=(t_1-t_0)/t_0$ and $k'_2=(t_2-t_0)/t_0$ are determined to be 1.07 and 1.86, respectively. Separation factor (α), which represents the chiral recognition ability of CSP, is obtained as $\alpha=(t_2-t_0)/(t_1-t_0)$. Here, α value is 1.75. Resolution factor (Rs) is obtained as

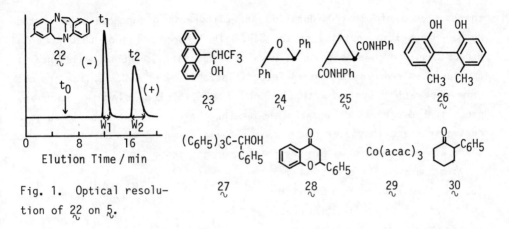

Fig. 1. Optical resolution of $\underset{\sim}{22}$ on $\underset{\sim}{5}$.

$Rs = 2(t_2 - t_1)/(W_1 + W_2)$ where W_1 and W_2 are peak widths. In Figure 1, Rs value is 3.09.

Cellulose Tris(4-substituted phenylcarbamate)

Table 1 shows the separation factor (α) in the resolution of racemic compounds ($\underset{\sim}{22}$-$\underset{\sim}{30}$) on cellulose trisphenylcarbamate derivatives ($\underset{\sim}{1}$-$\underset{\sim}{11}$) having a substituent at 4-position. In many cases, good chiral recognition was observed with alkyl-substituted derivatives or halogen-substituted derivatives including CF_3 group. Unsubstituted derivative $\underset{\sim}{6}$ did not exhibit the best chiral recognition for $\underset{\sim}{22}$-$\underset{\sim}{30}$. These results suggest that

TABLE 1

Separation factors (α) of racemates ($\underset{\sim}{22}$-$\underset{\sim}{30}$) on cellulose 4-substituted trisphenylcarbamate derivatives ($\underset{\sim}{1}$-$\underset{\sim}{11}$)

| | $\underset{\sim}{1}$ | $\underset{\sim}{2}$ | $\underset{\sim}{3}$ | $\underset{\sim}{4}$ | $\underset{\sim}{5}$ | $\underset{\sim}{6}$ | $\underset{\sim}{7}$ | $\underset{\sim}{8}$ | $\underset{\sim}{9}$ | $\underset{\sim}{10}$ | $\underset{\sim}{11}$ |
	CH_3O	CH_3	CH_3CH_2	$(CH_3)_2CH$	$(CH_3)_3C$	H	F	Cl	Br	CF_3	NO_2
$\underset{\sim}{22}$	~1	1.48	1.11	1.17	1.74	1.37	1.14	1.16	1.19	1.23	~1
$\underset{\sim}{23}$	1.35	1.52	1.57	1.59	1.75	1.45	1.26	1.29	1.29	1.30	~1
$\underset{\sim}{24}$	1.34	1.55	1.55	1.43	1.27	1.46	1.38	1.68	1.70	1.61	1.33
$\underset{\sim}{25}$	1.00	1.35	2.12	2.14	2.24	1.45	~1	1.44	1.17	1.22	~1
$\underset{\sim}{26}$	1.15	1.30	1.33	1.39	1.50	1.65	1.17	1.20	1.21	2.04	~1
$\underset{\sim}{27}$	1.00	1.37	1.59	1.47	1.36	1.22	1.64	1.95	1.95	1.48	1.00
$\underset{\sim}{28}$	~1	1.16	1.22	1.23	1.45	1.10	1.13	1.12	1.13	1.14	1.00
$\underset{\sim}{29}$	~1	1.75	1.76	2.46	2.50	1.24	1.53	1.46	1.79	2.06	~1
$\underset{\sim}{30}$	1.13	1.20	1.19	1.15	1.22	1.17	1.12	1.16	1.17	1.18	~1

the existence of electron-donating or -withdrawing groups can enhance the chiral recognition ability of the CSP. The introduction of CH_3O and NO_2 groups resulted in a decrease of α values, suggesting that the existence of a heteroatom reduces the chiral recognition ability of CSP. In the present chromatographic system with a nonpolar eluent, the polar interaction, probably hydrogen bond, between the urethane group of the phenylcarbamate derivatives and the polar group of a racemic compound is considered to be the most important force for the chiral recognition.

A schematic interaction is depicted in Figure 2. Both NH and CO of the carbamate group can interact with a solute through hydrogen bond. The existence of an electron-donating group like an alkyl group is likely to increase the electron density of the carbonyl oxygen, and the hydrogen bond on this oxygen may be strengthened. On the other

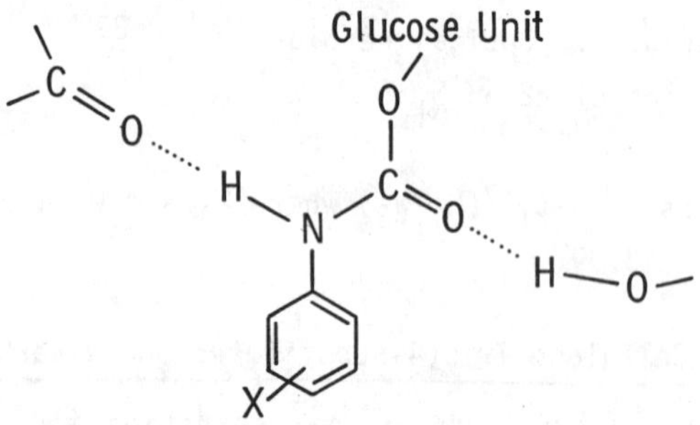

Fig. 2. Adsorbing site of cellulose tris(phenylcarbamate)

hand, the introduction of an electron-withdrawing group like halogen increases the acidity of NH proton. This has been confirmed by the downfield shift of NH resonance in ^{1}H NMR [3]. The increased acidity of NH group should increase the capability of the formation of the hydrogen bond on this group with an electron-donating group like carbonyl. Thus, the introduction of an alkyl or halogen group is expected to increase the chiral recognition ability of the cellulose tris(phenylcarbamate) derivatives. The polar substituents like CH_3O and NO_2 themselves can interact with a polar solute. Since these groups of the cellulose derivatives 1 and 11 exist far from the chiral glucose unit, the interaction on these groups can not discriminate enantiomers effectively. Therefore, the existence of these groups reduces the chiral recognition ability of CSP.

Among four alkyl-substituted carbamates 2-5, tert-butyl derivative exhibited remarkable resolving power. Five racemic compounds were best resolved on this column. This high resolving power may be due to both electronic and steric effects of tert-butyl group. The tert-butyl group may force a solute to interact with the carbamate group.

TABLE 2

Separation factors (α) of racemates (22-30) on cellulose tris(methyl-substituted phenylcarbamate)s (2, 12-17)

	2	12	13	14	15	16	17
	4-CH$_3$	3-CH$_3$	2-CH$_3$	3,4-(CH$_3$)$_2$	3,5-(CH$_3$)$_2$	2,6-(CH$_3$)$_2$	3,4,5-(CH$_3$)$_3$
22	1.48	1.45	~1	1.49	1.32	~1	1.48
23	1.52	1.56	1.10	2.13	2.59	1.17	1.77
24	1.55	1.28	1.35	1.13	1.68	~1	1.36
25	1.35	~1	~1	2.39	3.17	1.36	1.22
26	1.30	2.63	~1	1.87	1.83	1.34	2.09
27	1.37	1.45	~1	~1	1.34	~1	~1
28	1.16	1.14	~1	1.42	1.41		1.21
29	1.75	1.29	~1	1.32	~1	~1	~1
30	1.20	1.17	~1	1.20	1.15	~1	1.08

Cellulose Tris(methyl-substituted phenylcarbamate)

Table 2 shows the α values of 22-30 in the resolution on cellulose tris(methyl-substituted phenylcarbamate) derivatives 2 and 12-17. The chiral recognition by these derivatives depended very much on the number and position of the methyl group introduced. 3,5-Dimethyl derivative 15 exhibited remarkable resolving power for 23, 24, and 25. However, the derivatives 13 and 16 having a methyl group at 2- or 6-position possessed lower optical resolving ability. Most of cellulose tris(phenylcarbamate) derivatives formed a liquid crystal phase through hydrogen bond [5]. This means that the derivatives exist in an ordered structure on the surface of silica gel. Such an ordered structure may not be present in the case of 2- or 6-substituted derivatives because of steric hindrance of the methyl group which prevents the formation of hydrogen bond. We also prepared various chloro-substituted trisphenylcarbamate derivatives [3]. The influence of chloro-substituent is rather similar to that of methyl group and 3,5-dichloro derivative exhibited the most noticeable recognition.

Polysaccharide 3,5-Dimethylphenylcarbamate

Among many cellulose trisphenylcarbamate derivatives, 3,5-dimethyl derivatives 15 gave one of the most practically useful columns from the

TABLE 3

Optical resolution on 3,5-dimethylphenyl-
carbamate derivatives of polysaccharides

	15	18	19	20	21
	Cellulose	Amylose	Chitosan	Dextran	Xylan
22	1.32(+)	1.58(+)	~1(+)	1.26(+)	1.65(−)
24	1.68(−)	3.04(+)	1.10(+)	1.00	1.40(+)
25	3.17(+)	2.01(+)	1.33(−)	1.35(+)	~1(+)
26	1.83(−)	2.11(−)	1.17(−)	~1(+)	1.02(−)
27	1.34(+)	1.98(+)	1.27(+)	1.57(+)	1.23(+)
29	~1(+)	~1(−)	~1(+)	1.14(−)	2.57(+)

Optical rotation of first–eluted isomer is
shown in parenthese.

viewpoints of its easy availability, high chiral recognition ability and
stability. Then, other polysaccharides, amylose, chitosan, xylan and
dextran were converted to 3,5-dimethylphenylcarbamates (18-21) and their
optical resolving power as CSPs was estimated in the same way as described
above. The results are summarized in Table 3. The optical resolving
power of amylose derivative (18) is noteworthy. This derivative resolved
several racemic compounds more efficiently than the cellulose derivative
(15). The elution order of enantiomers was sometimes reversed as shown for
trans–stilbene oxide (24). The chitosan derivative (19) showed rather low
chiral recognition, although it has a structure similar to cellulose
derivative (15). The existence of a urea group (−NHCONH−) may disturb the
order structure which exists in the cellulose derivatives. The xylan
derivative (21) separated particularly effectively the enantiomers of 22
and 29. These compounds may be discriminated by adjacent carbamate groups
at 2– and 3–position of xylan.

Optical Resolution of Various Compounds

Besides the compounds 22-30, many racemic compounds were resolved on
these phenylcarbamate derivatives, particularly on 5, 15 and 18. The
examples of the compounds which were resolved on these three columns are
shown in Figure 3. The drugs (chloroquine and nicardipine) were resolved
only on 5. Amylose derivative 18 efficiently resolved β–blocker sotalol
which was difficult to separate with other columns. Cellulose derivative

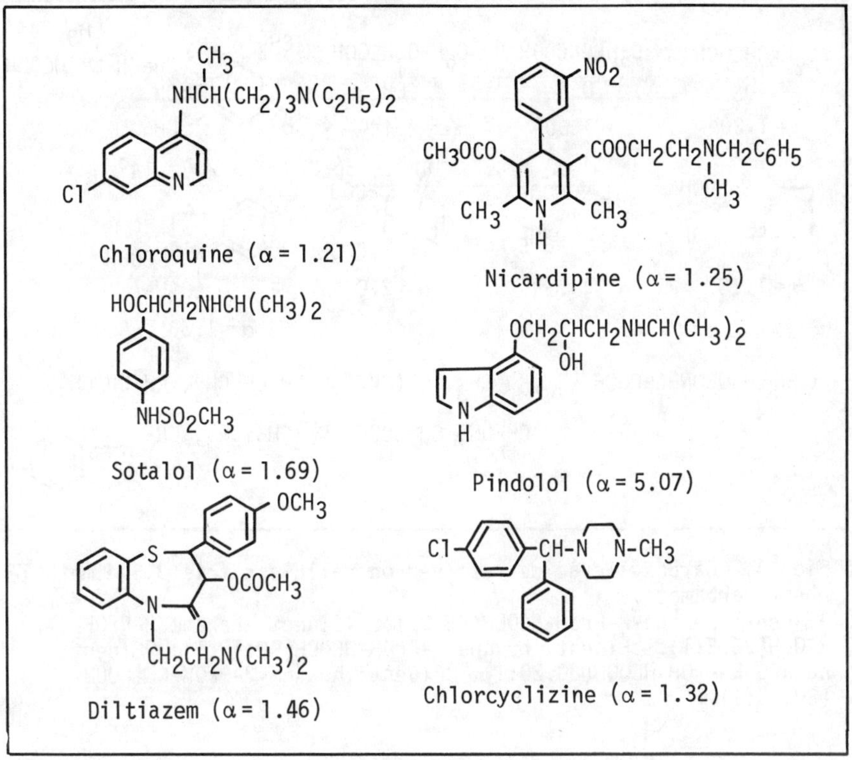

Fig. 3. Optical resolution of chloroquine and nicardipine on 5, pindolol, diltiazem and chlorcyclizine on 15 and sotalol on 18.

15 also resolved many drugs which include β-blocker (pindolol), chlorcyclizine and diltiazem [6,7]. This derivative 15 is also useful to separate directly racemic carboxylic acids as shown in Figure 4 [8].

So far, we examined the optical resolution of about 360 racemic compounds by 15 and about 230 (64 %) of them were resolved. The amylose derivative 18 resolved about 150 (56 %) of 270 racemic compounds examined. These percentages may be quite high compared with the percentages by other chiral columns so far developed.

The polysaccharide columns showed good durability toward the use for a long term and are applicable to preparative separation. Therefore, the columns will be widely used in many fields dealing with chiral compounds.

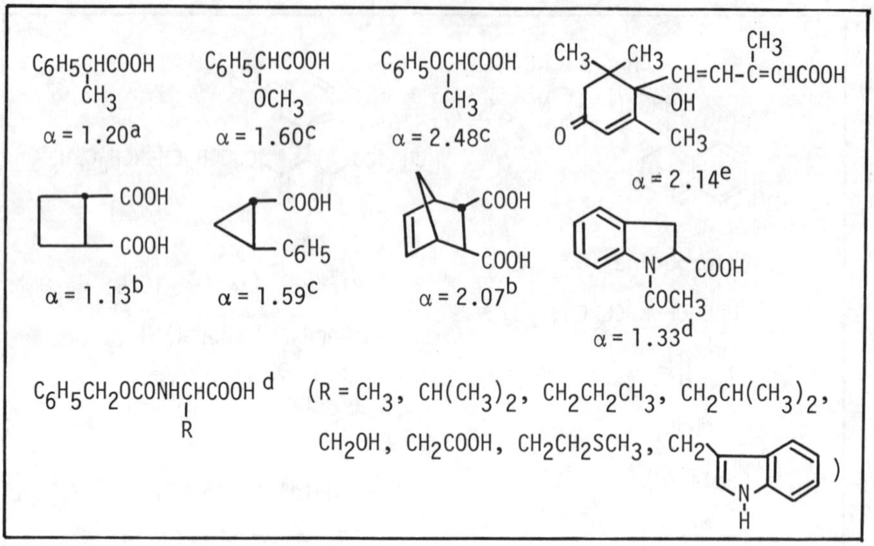

C6H5CHCOOH
|
CH3
α = 1.20ᵃ

C6H5CHCOOH
|
OCH3
α = 1.60ᶜ

C6H5OCHCOOH
|
CH3
α = 2.48ᶜ

[structure with CH3, CH3, OH, CH3, CH=CH-C=CHCOOH with CH3]
α = 2.14ᵉ

[cyclobutane dicarboxylic acid structure] COOH / COOH
α = 1.13ᵇ

[cyclopropane structure] COOH / C6H5
α = 1.59ᶜ

[norbornene dicarboxylic acid structure] COOH / COOH
α = 2.07ᵇ

[indoline structure with N–COCH3, COOH]
α = 1.33ᵈ

C6H5CH2OCONHCHCOOH ᵈ
|
R

(R = CH3, CH(CH3)2, CH2CH2CH3, CH2CH(CH3)2, CH2OH, CH2COOH, CH2CH2SCH3, CH2[indole])

Fig. 4. Carboxylic acids resolved on cellulose tris(3,5-dimethyl-phenylcarbamate).
ᵃEluent: hexane-2-PrOH-HCOOH(98:2:1). ᵇEluent: hexane-2-PrOH-HCOOH(95:5:1). ᶜEluent: hexane-2-PrOH-HCOOH(90:10:1). ᵈEluent: hexane-2-PrOH-HCOOH(80:20:1). ᵉEluent: hexane-2-PrOH-CF3COOH (80:20:1).

REFERENCES

1. Armstrong, D.W., Anal. Chem., 1987, 59, 84A ; Okamoto, Y. CHEMTECH, 1987, 176–181.
2. Okamoto, Y., Kawashima, M. and Hatada, K., Useful chiral packing materials for high-performance liquid chromatographic resolution of enantiomers: phenylcarbamates of polysaccharides coated on silica gel. J. Am. Chem. Soc., 1984, 106, 5357–5359.
3. Okamoto, Y., Kawashima, M. and Hatada, K., Chromatographic resolution XI. Controlled chiral recognition of cellulose triphenylcarbamate derivatives supported on silica gel. J. Chromatogr., 1986, 363, 173–186.
4. Okamoto, Y., Aburatani, R., Fukumoto, T. and Hatada, K., Useful chiral stationary phases for HPLC. Amylose tris(3,5-dimethylphenylcarbamate) and tris(3,5-dichlorophenylcarbamate) supported on silica gel. Chem. Lett., 1987, 1857–1860.
5. Vogt, U. and Zugenmair, P., Investigation on the lyotropic mesophase system cellulose tricarbanilate / ethyl methyl ketone. Makromol. Chem., Rapid Commun., 1983, 4, 759–765.
6. Okamoto, Y., Kawashima, M., Aburatani, R., Hatada, K., Nishiyama., T. and Masuda, M., Optical resolution of β-blockers by HPLC on cellulose triphenylcarbamate derivatives. Chem. Lett., 1986, 1237–1240.
7. Okamoto, Y., Aburatani, R., Hatano, K. and Hatada, K., Optical resolution of racemic drugs by chiral HPLC on cellulose and amylose tris(phenylcarbamate) derivatives. J. Liq. Chromatogr. in press.
8. Okamoto, Y., Aburatani, R., Kaida, Y. and Hatada, K., Direct optical resolution of carboxylic acids by chiral HPLC on tris(3,5-dimethyl-phenylcarbamate)s of cellulose and amylose. Chem. Lett., 1988, 1125–1128.

DISCUSSION

Q: R. Atalla (Inst. Paper Chem. USA)

The polysaccharides studied should have a wide range of ability of their derivatives to form liquid crystals; do they all have a correlation between liquid crystal characteristics and their chiral resolving ability ?

A: Chiral recognition abilities depend very much on coating conditions. This probably means that small change of the conformation of the derivatives may greatly influence the optical resolving power of the derivatives.

Q: D. G. Gray (McGill Univ.)

Is there any correlation between the effectiveness of the solvent used for coating and its ability to form a liquid crystalline phase ?

A: Yes, there is. Chiral recognition ability of chiral stationary phases depends very much on the conditions of coating.

Q: P. A. Williams (North East Wales Inst.)

Have you investigated the configuration of the cellulose derivatives adsorbed onto silica gel ? The configuration will depend on the solvent conditions and polymer molecular mass as well as the chemical nature of the polymer. A convenient method for studying the configuration of polymers at interfaces is ESR spectroscopy.

A: No, at present time, there is no good way. To understand the chiral recognition mechanism, we must determine exact structure.

Q: L. Vollbracht (AKZO)

Can amino acids be separated by your method?

A: No, amino acids are soluble only in water. In our systems, we must use non-polar eluents.

SUMMARY

given by P. Zugenmaier for Round Table I-2

The problem of organized molecular assemblies was discussed from different viewpoints. Question that arosed ; what structural models can be proposed, what are the interactions, the driving forces that specific structures are formed. Extensive answers have been provided by the panel and discussed with the audience.

ROUND TABLE II-1 <Bio-Tech>

"BIOTECHNOLOGY IN CELLULOSICS"

Chairman: G. O. Phillips
            (North East Wales Inst.)

            and

            A. Blažej
            (Slovak Tech. Univ.)

INTRODUCTORY REMARKS

given by G. O. Phillips

"Biotechnology" is a much hackneyed word, so we should be
more specific in our objectives for this session. May I
define the scope of this session as follows:
    Biosynthesis, Biodegradation of Cellulose, Lignin
    and various aggregates, usually referred to as
    lignocellulosics derived from wood and other
    naturally occurring materials.
Ever since I was a student in the University of Wales when
Stanley Peat and W. J. Whelan were beginning their classical
work on the enzymic synthesis and degradation of starch, I've
regarded enzymes with a mixture of awe and suspicion. Nevertheless, they have now become an indispensable part of the
armory of the carbohydrate industrial and research chemist.
However, enzymic advances in cellulosics have been slower
than in starch. This, of course, is related in the main to
the stubbornness of the cellulose structure to yield to
chemical, biological and physical treatment. But that is
exactly why too it is such an excellent material for
Nisshinbo Industries !
    Our specialists at this Round-Table Discussion will cover
the various problems which I can briefly summarize.
A. Pre-treatments
    Some mechanical, chemical or physical (such as γ-irradia-
tion) treatment is necessary to enhance the biodegradation,
by making the material more accessible.
B. Biodegradation of Wood
    The most efficient degrading enzymes are those derived
from white and brown rot fungi (basidiomycetes). The brown
rot fungi first attack and depolymerize cellulose and there-
after the hemicellulose, but leave the lignin largely
unaltered. The white rot fungi degrade both the lignin and
polysaccharides simultaneously. In whole wood the difficulty
is obtaining suitable observational techniques to track the
course of the degradation.
C. Biodegradation-Biosynthesis of Cellulose
    Crystalline cellulose is hydrolyzed by a 2 or 3-step
reaction sequence involving not only hydrolytic enzymes, but
also oxide-reductive enzymes. Very special qualities are
called for in these enzymes. As we will hear the Acetobacter
aceti bacterium can effectively produce cellulosic material,
with considerable applicational potential.
D. Biodegradation of Lignin
    While lignin is resistant to most microorganisms, the
oxidative degradation by white rot fungi systems has now
yielded a great deal of scientific information, which is now
contributing information about the structure of lignin within
wood itself.
    There are many questions to answer. My cochairman, Dr. A.
Blažej of the Slovak Technical University will now briefly
outline some of these.

FURTHER REMARKS

given by A. Blažej

Bioconversion of lignocellulosics has for next future a great importance for the practice.

We have at present relatively good fundamental knowledge on enzymes degrading of wood and non-wood materials (straw, baggase etc.) as a whole and also good knowledge on respective groups of enzymes degrading main component of wood e.g. cellulase, xylanase, ligninase from the point of view of the chemical structure, enzymology, etc. At present some pilot productions or small scale productions of cellulases, mainly from Trichoderma species are in operation. The most attractive fungi is <u>Trichoderma reesei</u>.

Potential possibilities for next future promise the genetically engineered microorganisms which can increase the productivity of enzymes production for lignocellulose degradation. The new technique of molecular genetics provides a new stimulation and challenge for commercialization of the fundamental research in this area.

What I would like to stress is that polysaccharides hydrolyzed by enzyme systems of microorganisms, and the same holds in the case of the ligninase ones, show <u>unusual complexicity</u> in terms of the number of enzymes produced by microorganisms. For example, we have studied in Bratislava (Czechoslovakia) with my coworkers Dr. Biely and Dr. Markovič enzymes from Trichoderma reesei using two-dimensional separation of enzyme components by electrophoresis and isoelectric focusing followed by detection of enzyme activities with chromogenic and fluorogenic substrates. We have isolated and characterized 18 different glycanases of Trichoderma reesei with respect to cellulose and xylan degradation. Four different types of glycanases related to wood polysaccharides degradation were identified:

a) specific endo-1,4-β-glucanase which seems to be identical with the so-called cellobiohydrolase II (CHB II):

b) specific endo-1,4-β-xylanase, presumably two types

c) nonspecific endo-1,4-β-glucanase (hydrolyzing also xylan and β-lactosides as the agluconic bond)

d) cellobiohydrolase I (exo-1,4-β-glucanase) hydrolyzing β-lactosides and exhibiting no endoglucanase, endoxylanase or exoxylanase activity.

With the exception of one specific endo-1,4-β-xylanase, all other glycanases occur in multiple forms with different isoelectric points.

An understanding of the nature of various cellulolytic, xylanolytic and ligninolytic enzymes and their enzyme activities on different substrates of lignocellulosics is important for theoretical and practical reasons. In spite of intensive research work in many laboratories there is still considerable uncertainty about the substrate specificity of glycanases and ligninases with respect to cellulose, xylan and lignin degradation.

The question of the specificity has not been solved even

for the most extensively studied enzymes degrading wood component.

The reason lies in the difficulties in purifying some of the complex enzyme systems due to the multiciplicity of their forms and their very similar physicochemical properties.

I do personally believe that the specificity of the polysaccharides and lignin degrading enzymes is given by stereospecificity of polysaccharides and lignin and their conformation structure. For example in the case of polysaccharides the conformation structure of glycosidic bonds is different for respective polysaccharides and the electrospherical microenvironment of respective glycosidic bond in polysaccharide chain is different, too. Enzyme specificities of cellulolytic enzymes are given with high probability by different stereospecificity of glycosidic bonds due to theirs different conformation of individual polysaccharides.

For industrial treatment of lignocellulosics one of the best processes is offered by biotechnology and in next future it is a very important area to produce higher added value products based on biochemical treatment of wood and non-wood lignocellulosics.

Production and Application of Bacterial Cellulose

SHIGERU YAMANAKA

Central Research Laboratories of Ajinomoto Co., Inc.
1-1 Suzuki-cho, Kawasaki-ku, Kawasaki, Kanagawa 210, Japan

ABSTRACT

Acetobacter aceti, an acetic acid bacterium, is known to produce a cellulosic material. The latter is produced as a gel-like substance on the surface of the culture medium in a static culture. This bacterial cellulose (=BC) has been studied by many researchers from various points of view, i.e., as to the supermolecular structure of the pellicle, fermentative production, the biogenesis of cellulose, etc.

We have been interested in the mechanical properties of BC and its industrial application. We have already reported the following; a sheet-like material derived from gel-like BC produced under static culture conditions has remarkable mechanical strength, viz. the Young's modulus is as high as > 15 GPa. This material is expected to be a new industrial material suitable for transducer diaphragms, for example. Moreover, disintegrated BC seems to be useful for reinforcing various types of papers.

The above findings prompted us to investigate effective production methods for this material. The most potent strain, Acetobacter aceti AJ 12368, was selected under static conditions. This strain segregated into two types, one forming rough type colonies (R type) and the other smooth type colonies (S type). Under static conditions, the R type produced more BC than the S type. In a search for effective components, phytic acid was found to promote the production of BC. For increased BC production, aerated or agitated culture conditions were investigated in comparison with static culture conditions. However, static culture conditions seem to be superior to aerated or agitated culture conditions from an industrial point of view.

INTRODUCTION

Some strains of <u>Acetobacter</u> <u>aceti</u> produce an extracellular gel-like material or pellicle, which comprises of a random assembly of cellulose ribbons composed of a number of microfibrils (bacterial cellulose = BC). We found that such a specific supermolecular structure gave interesting mechanical properties, from the view point of new materials, in collaboration with researchers at SONY Co., Tokyo, and the Research Institute for Polymers and Textiles, Tsukuba [1,2]. On drying of this gel-like material a sheet with a high Young's modulus was obtained. This sheet is expected to be a new industrial material suitable for transducer diaphragms. The high Young's modulus of this material was considered to be due to its supermolecular structure. Scanning electron microscopy of the sheet revealed that it comprised of a pile of very thin layers, which consist of ribbons of the native pellicle. These layered ribbons were thought to be bound through strong hydrogen-bonds. Fragments of BC were reported to be obtained on mechanical disintegration in water. They consist of entangled ribbons, and are expected to be useful for reinforcing various types of papers and for the processing of other fibrous materials into a form of paper, a small amount of the disintegrated BC being added.

With this background, we studied the fermentative production of this material. Much research has been performed on the production of BC on a laboratory scale, but mainly under static culture conditions. Culture medium saccharides have been used as carbon sources, and yeast extract and $(NH_4)_2SO_4$ as nitrogen sources. After the selection of a potent BC producer, we searched for effective substances in the culture medium. Cellulose is produced at the surface of a liquid culture medium in a static culture because oxygen is needed for its production. As aerated or agitated cultivation conditions supply a higher quantity of oxygen, BC production was examined with various culture methods.

METHODS AND MATERIALS

1. Microorganism
 <u>Acetobacter</u> <u>aceti</u> AJ12368 was used throughout this work.
2. Culture medium
 The standard culture medium used was composed of 50g sucrose, 5g yeast extract, 5g $(NH_4)_2SO_4$, 3g KH_2PO_4 and 0.05g $MgSO_4$ $7H_2O$, dissolved in one liter of water (pH5.0).

3. Culture conditions
 The microorganism was precultured in a 500 ml flask containing 400 ml of the culture medium at 30° C for 2 days with shaking (120 reciprocal units per min., 7cm stroke). The production of cellulose was examined in various types of vessel, 1 % of precultured broth being innoculated into the medium, under static, vibrated or agitated conditions at 30°C.
1. Weighing of cellulose
The gel-like cellulose was weighed in the following ways;
(1) Pressed wet weight: Pellicles were harvested, washed in

running water, boiled in 10 volumes of a 2%(w/v) NaOH solution
for 1 hour and finally thoroughly washed in running water
again. The washed pellicles were pressed between two plates at
0.3 kg/cm^2 and the weight of the resultant pressed material was
taken as the pressed wet weight.
2) Dry weight: Thoroughly washed pellicles, as above, were
dried in an oven at 105°C to a constant weight.

RESULTS AND DISCUSSION

1. Screening of effective substances for BC production

Acetobacter aceti AJ 12368 was used for the production of
BC. This strain segregated into two distinctive types of
colonies on agar medium, rough (R strain) and smooth (S
strain) type colonies, as shown in Fig. 1. Both strains
produced cellulose, but the R strain produced a much higher
quantity (13 g pressed wet weight per plate; plate area, 154
cm^2, depth of medium, 6mm) than S strain (2 g, ibid.) in a
static culture for 7 days, respectively. Therefore, the R
strain was used for further experiments.

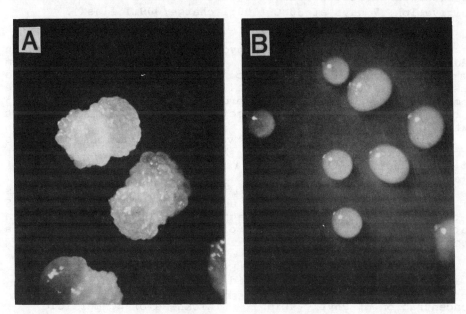

Figure 1. Colony types of AJ12368. A:Rough type, B:Smooth type.

Effective substances for BC production were searched for,
with particular attention to yeast extract, an organic nitrogen
source of the medium. Various such components were compared
for BC production as to the thickness of the resultant
cellulosic pellicles (Table 1). Yeast extract (0.5%) and a
soybean hydrolysate (0.25%) gave the highest accumulation,
followed by corn steep liquor and Soytone(Difco, U.S.A.).
Casamino acid (Difco), Polypeptone (Daigo, Japan) and Casein

(Difco) whose major constituents are amino acids or proteins, were shown to be poor medium sources, which suggests that factors other than amino acids or proteins are needed for elevated BC production.

TABLE 1
Effects of various organic nitrogen sources on BC production

Source	Concentration (%)	Thickness of BC pellicle (mm)
Yeast extract (Difco)	0.5	9
Casamino acid (Difco) (vitamin-free)	0.5	2
Polypeptone (Daigo)	0.5	3
Soytone (Difco)	1.0	5
Casein (Difco)	0.5	2
Soybean hydrolysate	0.25	9
Corn steep liquor	1.0	6

Cultivation: 7 days at 30°C (100ml charge/300ml flask).

Yeast extract and soybean hydrolysate commonly contain, in addition to amino acid sources, vitamins B_2 and B_6, calcium pantothenate, biotin, niacin, inositol, phytic acid (hexaphosphate ester of inositol), choline, etc. These compounds were added to the BC production medium containing 0.5 % casamino acids as the amino acid component instead of yeast extract.

Almost no effect was observed for the various compounds except phytic acid, phytic acid showing a high promoting effect of BC production. Phytic acid is known to be a constituent of plant seeds.

BC production with the addition of various concentrations of phytic acid was investigated using the medium containing 0.5 % casamino acids as the amino acid component (Table 2). This table shows that the addition of 0.2 g/l of phytic acid was enough.

TABLE 2
Effect of phytic acid on BC production

Concentration of Phytic acid (mg/l)	Thickness of BC pellicle (mm)
0	3
20	7
200	10
2,000	11

Cultivation: 1 week at 30°C (100ml charge/300ml flask).

On the basis of the above findings, the following culture

medium was established; 50 g/l sucrose, 5 g/l amino acid mixture, 0.2 g/l phytic acid, 3 g/l KH_2PO_4, 2.4 g/l $MgSO_47H_2O$, and 1 g/l $(NH_4)_2SO_4$ (pH 5.0).

2. Culture methods

Various culture methods were compared as to BC production. The synthesis of BC has mainly been studied in static cultures [3,4], little research having been reportedly done on stirred cultures [4]. Cellulosic pellicles are produced on the surface of the culture medium, i.e., the liquid-air interface, in static cultures, and, therefore, oxygen supply seems to be the limiting factor. We considered that an elevated oxygen supply would increase the production and so investigated the following two methods: firstly the oxygen concentration in the gas phase was increased to above 21 % by the addition of oxygen gas to the air in a static culture and, secondly, a vibration or agitation culture was performed.

In a static culture with 21 % oxygen (air), BC was produced at the highest rate and a higher concentration decreased its production, presumably because of the toxic effect of oxygen (Table 3).

TABLE 3
Effect of the oxygen concentration on BC production.

Oxygen concentration (%)	Thickness of BC pellicle (mm)	
	7th day	14th day
21	9	16
30	3	12
50	2	6
90	2	3

300 ml tall form beaker containing 200 ml of medium.
Yeast extract was used as an organic nitrogen source.

In vibration and agitation cultures, the morphology of BC varied with the conditions and culture time; pellet-form, filamentous or irregular masses attached to the wall of the vessels, etc. In the case of a conventional small glass jar fermenter, harvesting of the gel-like substance was not easy because some of it became attached to the impeller or wall of the fermenter, etc. In a static culture, approximately 8 g/l of dry BC was accumulated, while in a stirred culture the concentration of the product varied between 2-3 g/l, according to the culture conditions as shown in Table 4 as examples. Though static cultures and stirred cultures seemed to be quite difficult to compare directly, static cultures have given the highest concentrations so far.

TABLE 4
Accumulation of BC in static cultures and agitated cultures.

Conditions	BC produced (g/l, 3days)
Static[*1]	8.4
Vibrated[*2]	2.5
Agitated[*3]	
100 rpm	1.7
300 rpm	2.0

[*1] 100ml charge/154cm^2 petri dish.
[*2] Stroke, 0.055mm; 330Hz (petri dish culture as above).
[*3] Petri dish culture as above.
Amino acids and phytic acid were used instead of yeast extract.

CONCLUSION

For BC production, phytic acid was found to be a promoting factor. Static cultures gave the highest concentrations of BC, in comparison with stirred cultures.

REFERENCES

1. Yamanaka, S., Watanabe, K., Kitamura, N., Iguchi, M., Mitsuhashi, S., Nishi, Y. and Uryu, M., Superstructure of bacterial cellulose gel. Polymer Prepr. Japan 1987, 36, No.4, 951.
2. Nishi, Y., Uryu, M., Yamanaka, S., Watanabe, K., Kitamura, N., Iguchi, M. and Mitsuhashi, S., Superstructure of bacterial cellulose gel. Polymer Prepr. Japan 1987, 36, No.4, 952.
3. Dudman, W.F., Cellulose production by Acetobacter acetigenum in defined medium. J. gen. Microbiol. 1959, 21, 327-337.
4. Shramm, M. and Hestrin, S., Factors affecting production of cellulose at the air/liquid interface of a culture of Acetobacter xylinum. J. gen. Microbiol. 1954, 11, 123-129.

DISCUSSION

Q: K. Nishinari (Nat. Food Res. Inst. Japan)
(1) On the effect of pressure on the tensile strength, you mentioned that tensile strength decreased with increasing pressure. Is it due to macroscopic or microscopic defect ?
(2) In relation to this, does the density of the sample change ?

A: (1) I do not know if it is due to macroscopic or microscopic defect. Anyhow some part of cellulosic pellicle was speculated to be injured with increasing pressure.
(2) No, it does not change.

Q: G. Franz (Univ. Regensburg)
    Chemical purity of bacterial cellulose : no contamination
of other type glucans i.e. β-1,3 or β-1,2 ? If this is the
case the material should be ideal for hemodialysis membranes.

A: From samples just after being harvested from cultured
broth it contains other type glucan. It can be purified on
alkaline treatment and so on to more than 99%.

Q: A. Blažej (Slovak Tech. Univ.)
    Did you investigate to prepare fibers from BC ? Is it
possible to formulate fibrous material from BC ?

A: Unfortunately it is not possible to formulate fibrous
material from BC at present. This is a future problem.

Q: R. M. Brown, Jr. (Univ. Texas)
    (1) Does multilayer structure arise because of high den-
sity pellicle which sinks, thus initiating a new pellicle ?
    (2) (Comment) Tiu lapse video shows separation of rib-
bons during synthesis. Therefore no true "branches" connected
by glycosidic bonds are produced. Only physical associations
prevail in this case.

A: (1) Yes, it does.
    (2) My speculation for branching system is based on total
observation of electron micrograph of bacterial cellulose. So
it is necessary to have more evidence to confirm these points
with certainty.

CELLULASE PRODUCTION AND BIOMASS CONVERSION WITH IMMOBILIZED AND CULTURED CELLS

ISAO KAETSU
Research Institute for Science and Technology,
Kinki University,
Kowakae 3-4-1, Higashi-Osaka, 577 Japan

ABSTRACT

Biomass conversion process consisting of pretreatment of cellulosic wastes, enzyme production with immobilized cellulase producing fungi and continuous saccharification, was studied. Effect of electron beam irradiation on crushing and saccharification was compared in various wastes. The promotive effect of radiation was remarkable in complexed lignin-cellulose-hemi cellulose containing wastes and it decreased time and energy of crushing. In order to carry out the saccharification advantageously, immobilization and continuous culture of the immobilized fungi for enhanced cellulase production were studied. Trichoderma reesei and Sporotrichum cellulophirum were immobilized in fibrous networked supports. The immobilized fungi showed an optimum cellulase production under suitable immobilization and culture conditions. The continuous culture of immobilized fungi gave the results of relatively constant FPA values for 20-30 days. The continuous saccharification was carried out for longer than one month with a constant glucose formation.

INTRODUCTION

The author's group has studied and developed a novel immobilization method of biofunctional components by a physical entrapping by means of radiation polymerization at relatively low temperatures. This method has been applied to the immobilization of enzymes, antibodies, drugs and microbial cells(1-5). One of the applications has been done on the immobilization of enzyme producing fungus and yeast in relation to a development of biomass conversion process(6-9). The process consisting of pretreatment of cellulosic wastes, immobilization and culture of fungus and saccharification of treated waste with cellulase produced in the immobilized cell culture was studied. The pretreatment was carried out by a combination of radiation degradation and mechanical crushing of the cellulosic wastes.

Few reports have described differences due to the effect of radiation pretreatment of various cellulosic wastes, in immobilization of fungus and yeast, and in culture of free and immobilized fungi. In this report, radiation effects in the pretreatment and immobilization and cellulase production of fungi were mainly studied in relation to the biomass conversion of cellulosic wastes.

RESULTS AND DISCUSSION

1. Radiation Effect on Pretreatment of Wastes (10-11)
   It is known that pretreatment is necessary to crush a

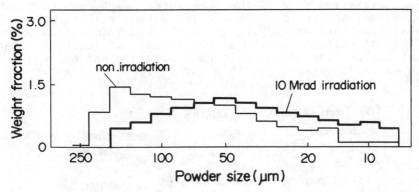

Fig.1. Effect of irradiation on size distribution of chaff irradiated and crushed by turbo-mill.

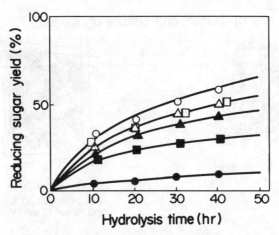

Fig.2. Effect of irradiation on saccharification of wastes
     ⊙,100Mrad irradiated chaff, ●,non-irradiated chaff,
     ⊡,100Mrad irradiated bagasse, ■,non-irradiated bagasse,
     △,100Mrad irradiated waste paper,▲,non-irradiated waste
   paper.

cellulosic waste and expose the cellulose for an effective
enzymatic saccharification.Effect of electron beam irradiation
on crushing and saccharification of various wastes was investigated. It was found that the irradiation was effective to destroy
a complexed structure of lignin,hemicellulose and cellulose,and
to decrease the mechanical strength. As a result, the irradiation caused a remarkable promotion of crushing to save the time
and energy of it. Figure 1 shows a shift of powder size distribution of irradiated chaff. However,the radiation effect
produces a decrease of chaff size. However, the irradiation
effect was not so clearly observed in the wastes of less
complexed structure and poor lignin content such as waste
paper as shown in Fig.2. It is suggested that the effect
strongly depends on the kind of waste.

2.Immobilization and Culture of Fungi
An immobilization of cellulase producing fungi such as Trichoderma reesei and Sporotrichum cellulophirum was investigated.
It was found that entrapping of fungi with hydrophilic polymer
gel was not suitable to the immobilization of yeast, but the
use of hollow fibrous supports such as gauze and non-woven
cloth was advantageous for this purpose. Fogure 3 shows

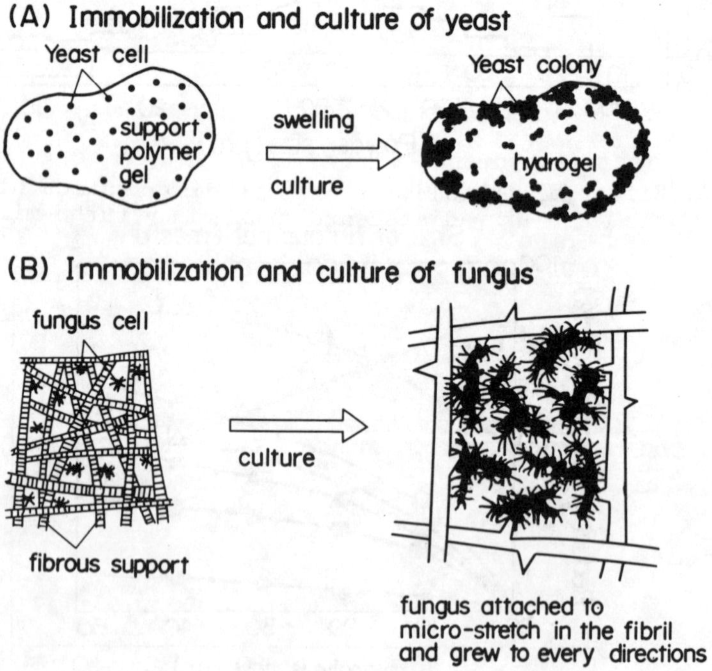

Fig.3. Model scheme for immobilization and culture of
yeast and fungus.

a model scheme for the principle of immobilization of fungus
and yeast. Fungi stretch the arms of mycelium and attach to a
micro-stretch on a surface of fibril of support by twining.
Then it spreads quickly into the hollow space of fibril network. It continues to grow until the whole space is occupied
and covered. On the other hand,yeast can grow inside matrix
of hydrogel to form colonies in the entrapped state.

Effects of various immobilization and culture conditions
on the growth and cellulase production measured by FPA(Filter
Paper Activity) were investigated. For example, an increase of
hydrophilicity of polymer support expressed by a water content
caused a decrease of CMC activity(carboxy methyl cellulose

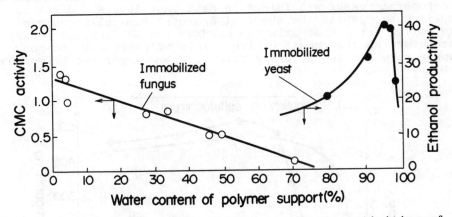

Fig.4. Effect of hydrophilicity of support on activities of
immobilized yeast and fungus.

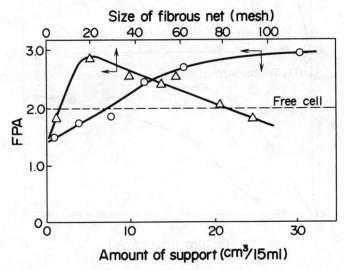

Fig.5. Effects of network size and packing amount of support
on FPA of immobilized and cultured Trichoderma reesei.

activity) of the culture product soup in the immobilized fungus as shown in Fig.4,while the immobilized and cultured yeast showed an increased ethanol fermentation activity with the increase of hydrophilicity of support. The properties of fibrous support affected on FPA values remarkably. According to the results of Fig.5, FPA in the immobilized and cultured Trichoderma reesei had a maximum at an optimum mesh size of fibril network in the support. It increased with increasing amount of fibrous support packed in the incubator. As shown in Fig.5, the obtained FPA values reached higher values under suitable conditions than that in a free cell culture.Geometrical form and structure of support in the incubator gave a large effect on the culture result. The fibrous sheet of immobilized Trichoderma reesei was folded to have many layers and set in the incubator,while the sheet of Sporotrichum cellulophirum was rolled arround an axis and used in the cylindrical rotary incubator. As the fungi require lots of oxygen,the oxygen supply rate and the stirring rate gave an important effect on

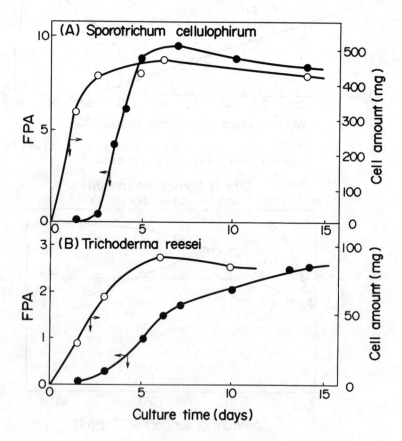

Fig.6. Relation between cell growth and cellulase production in the course of cultures of immobilized fungi.

the FPA results. The amount of cellulose powder added as an
inducer of cellulase production was also one of the important
factors to affect the cell culture. The cell growth of immobilized fungi was promoted with the increase of cellulose addition,
but the FPA value showed a maximum at a certain cellulose
concentration. As shown in Fig.6,a quick cell growth first
occurred and then a cellulase production was initiated with a
small delay from the cell growth in the culture of immobilized
fungi. The immobilized Sporotrichum cellulophirum showed much
quicker growth and cellulase release with higher FPA value
than in the immobilized Trichoderma reesei.But, the cell amount
and FPA of the immobilized Sporotrichum cell reached the

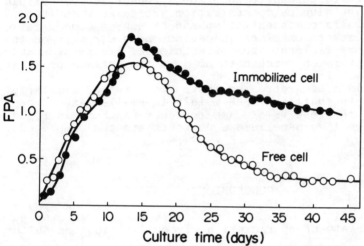

Fig.7. Change in FPA in the cultures of free Trichoderma reesei
and immobilized Trichoderma reesei.

saturation soon owing to excess increase of cell. It was found
that the immobilized Trichoderma reesei kept the considerably
enhanced FPA values during the continuous culture longer than
40 days.On the other hand,the FPA in the free cell culture in
a batch system decreased relatively quicker as shown in Fig.7.
The product soup in the culture was analyzed for the cellulase
compositions. It was found that the cellulase composition in
the soup of immobilized fungus culture was different from that
of free fungus culture. The cellulases from the immobilized
Sporotrichum cellulophirum contained a richer amount of
cellobiase than that from the free Sporotrichum cell.
     It was also found that the FPA increased again after the
gradual decrease in a long continuous culture by changing and
renewing the support with the immobilized fungus partly and
at intervals.The control of the most suitable amount of grown
cells in the support and of the most suitable packing ratio
of the support in the incubator as well as its geometrical
form and structure were the most important factors in the
biological engineering of such a culture system with the
immobilized fungus.
     Conclusively,it can be said that a relatively long
continuous cell culture and an enhanced cellulase production

are possible with an immobilized fungus in   fibrous supports
such as gauze and non-woven cloth. Detailed study on the
engineering will be reported in the near future.

3.Saccharification of Wastes with Cellulase
The pretreated cellulosic wastes such as chaff and bagasse
were mixed with water to make a slurry. Then the continuous
saccharification was carried out in   50 liter reactors with
cellulase solution. One of the typical results is shown in
Fig.8. According to this figure, almost constant concentration
of glucose was observed in the saccharification product of
pretreated chaff for a considerable period. The saturated
concentration corresponded to about 50 % conversion of cellulose
in the waste. The glucose concentration increased with the
increase of substrate concentration and of enzyme concentration.
The saturated conversion of cellulose depended strongly on the
conditions of pretreatment and powder size of pretreated waste.
The maximum conversion reached to 85 % at the optimum pretreatment conditions.
     Various kinds of pretreated wastes were tested similarly
for continuous saccharifications mainly in smaller laboratory
scale and gave the results of same tendency as those in Fig.2
according to the lignin-cellulose contents and the complexed
structure.

                         CONCLUSION

A biomass conversion process of cellulosic wastes was studied
including pretreatment by radiation, immobilization and culture

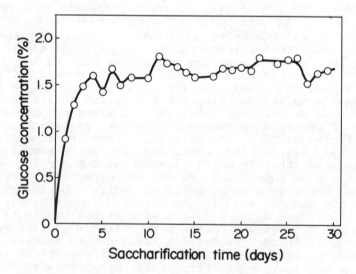

Fig.8. Result of a continuous saccharification of irradiated
       (10Mrad) and crushed chaff(10 %) with cellulase(1 %)
       at 40°C.

of fungi for cellulase production and saccharification by cellu-
lase. Irradiation|by electron beam caused the promotion of
crushing of waste into smaller size distribution and save of
crushing energy. But,the effect was observed only for the wastes
having complexed structure of lignin and cellulose.

Immobilization of fungus on a hollow fibrous materials
caused the enhanced growth and cellulase production in a continuous culture of fungi. Effects of various factors in the immobilization and cell culture on the FPA values were investigated
to find the optimum conditions. It was found that the controls
of network size of support,of packing amount and form of support
in the incubator,and of increase of cell amount in the support
were very important for the enhanced cellulase production.

Continuous long culture of immobilized fungi and continuous long saccharification of pretreated wastes with the cellulase were carried out in a bench scale and gave the products
stably for the considerably long period.

REFERENCES

1. Kaetsu, I.,Enzyme immobilization by radiation-induced polymerization of 2-hydroxyethyl methacrylate at low temperatures.
Biotechnol. Bioeng.,1979, 21,847.
2. Kumakura,M.,Kaetsu,I.,Suzuki,M.,and Adachi,S.,Immobilization
of antibodies and enzyme-labelled antibodies by radiation
polymerization. Applied Biochem.Biotechnol.,1983,8,87.
3. Yoshida,M.,Kumakura,M.,and Kaetsu,I.,Controlled release of
biofunctional substances by radiation-induced polymerization
(1)Release of potassium chloride by polymerization of various
vinyl monomers. Polymer, 1978, 19, 1375
4. Kumakura,M.,Yoshida,M.,and Kaetsu,I.,Immobilization of
Streptomyces Phaerochromogenes by radiation-induced polymerization of glass-forming monomers. Biotechnol. Bioeng.,1979
21, 679.
5. Kaetsu,I.,Kumakura,M.,Fujimura,T.,Yoshida,M.,Asano,M.,Kasai,
N.,and Tamada,M.,Studies on the immobilization of biofunctional
components by radiation polymerization and their applications.
Radiat.Phys.Chem.,1986, 27, 245.
6. Fujimura,T.,and Kaetsu,I., Immobilization of yeast cells by
radiation-induced polymerization. Z.Naturforsch.,1981,37c,102.
7. Fujimura,T.,and Kaetsu,I.,Growth of yeast cells immobilized
with porous swelling carriers by radiation polymerization.
J.Applied Biochem.Biotechnol.,1983,8, 145.
8. Kumakura,M.,and Kaetsu,I.,Immobilization of Trichoderma reesei
cell by radiation polymerization. Eur.Applied Microb.Biotechnol.
1983, 17, 197
9. Kumakura,M.,and Kaetsu,I.,Cellulase production from immobilized
growing cell composite prepared by radiation polymerization.
Biotechnol. Bioeng., 1984,116, 345.
10. Kumakura,M.,and Kaetsu,I.,Radiation-induced decomposition
and the enzymatic hydrolysis of cellulose.Biotechnol.Bioeng.
1978,20, 1809.
11. Kumakura,M.,Kojima,T.,and Kaetsu,I.,Pretreatment of lignocellulosic wastes by combination of irradiation and mechanical
crushing.Biomass,1982, 2, 299.

DISCUSSION

Q: J. Hayashi (Hokkaido Univ.)

(1) Why did not show the effect waste newspapers containing lot of lignin, though you mentioned that the irradiation pretreatment is effective for cellulosic materials containing lignin ?

(2) I think it is widely recognized that the most important thing for the pretreatments is to make porous structure for penetrating of enzyme and not to decrease crystallinity and to make small size powder. Some effective pretreatments are accompanied by decreasing of the crystallinity, however, it is not essential. What do you think about it ?

A: Radiation degradation on cellulose causes the decrease of mechanical strength of the composite (consisting of lignin-cellulose-hemicellulose) and so causes collapse of the original composite structure. This makes cellulose nakedly opened (exposed) so as to be able to contact with enzyme easily. I would say that the radiation is effective only on the materials having a rigid complexed structure of lignin-cellulose-hemicellulose. The newspaper seems to loose the original complexed structure, though it contains lignin still in another secondary form. I think that the total surface area for the contact with enzyme is essentially important. Therefore, particle size, porosity in the particle, and crystallinity of the particle are all important factors closely related to the effective surface area for enzyme reaction.

Q: R. Atalla (Inst. Paper Chem. USA)

Is it possible that a primary effect of irradiation is to cause crosslinking, making the fibers more brittle, and therefore, more easily fragmented, giving smaller particles ?

A: Cellulose is usually degraded by irradiation. (Each kind of polymer has a pattern to be degraded or crosslinked predominantly.) We have never observed a crosslinking of cellulose by irradiation. Brittleness can be attributed to the decrease of molecular weight due to chain scission with the increase of irradiation doses. In order to accelerate the degradation (to decrease the necessary dose), the additives, such as chlorine-containing compound may be effective.

Q: S. Hossain (Kimberly-Clark Co.)

I wonder if a preliminary energy balance, i.e. energy input (irradiation, crushing etc.) and the energy output (ethanol production) has been done. If so, I would be interested in the result.

A: According to our preliminary rough estimation, energy input was expected to be considerablly smaller than the output. However, the project group is now carrying out the evaluation work for the process and it would be summarized within a year in the future. I would send you a further information, when the work would be finished.

MECHANISMS OF THE FUNGAL AND ENZYMATIC DEGRADATION OF LIGNIN

TAKAYOSHI HIGUCHI

Wood Research Institute, Kyoto University, Uji, Kyoto 611, Japan

ABSTRACT

Lignin degradation by lignin-degrading basidiomycetes is explained based on the reaction mechanisms for cleavages of side chains and aromatic rings of arylglycerol-β-aryl ether substructure model compounds, and of synthetic lignin (DHP). Tracer studies using ^{2}H-, ^{13}C- and ^{18}O-labeled substrates with ^{18}O$_2$ and H$_2$^{18}O indicated that both lignin peroxidase and laccase of the lignin-degrading fungi Phanerochaete chrysosporium and Coriolus versicolor are initial enzymes in degradation of lignin. These enzymes catalyze the one-electron oxidation of nonphenolic and phenolic moieties in lignin, resulting in cleavages of side chains and aromatic rings. The roles of these enzymes in lignin biodegradation, and applications of biodegradation systems for delignification, are discussed.

INTRODUCTION

Recent progress in the chemistry and biochemistry of lignin biodegradation (1-4) suggests that degradative reactions of lignin mediated by lignin-degrading microorganisms and their enzymes might be applied for biomechanical pulping, biobleaching of pulp, and for treating waste waters in the pulp and paper industries. To establish the biotechnical (bioengineering) utilization of lignocellulosic materials, in which polysaccharides are intimately associated with lignin, the mechanism of lignin biodegradation should first be elucidated.

Lignin biodegradation has been studied through two complementary approaches: i) Degradation of polymeric lignin (5) such as synthetic lignin (DHP), milled wood lignin (MWL), and protolignin in xylem cell walls by lignin-degrading basidiomycetes (white-rot fungi). By the early 1980s analyses of decayed wood lignins by white-rot basidiomycetes had provided a general view or outline of lignin biodegradation such as cleavage patterns of propyl side chains and aromatic rings of lignin by the fungi. ii) However, specific degradative reactions involved in lignin biodegradation by the fungi have only been elucidated by using lignin substructure model compounds (2,6) as substrate for fungal degradation, because of great complexity and heterogeneity of the lignin polymer which consists of phenylpropane units connected via many C-C and C-O-C linkages. By the early 1980s many degradative reactions by the fungi, such as C$_\alpha$-C$_\beta$ cleavage of propyl side chains and cleavage of the β-O-4 bond were elucidated by studies using lignin substructure model compounds (1-3, 6).

Whereas aromatic ring cleavage products of lignin substructure model compounds by white-rot basidiomycetes were not identified until 1985, earlier studies (5) of polymeric lignin degradation by the fungi suggested involvement of ring cleavage reactions, based on chemical and spectroscopic analyses of degraded lignins. In 1985, we (7,8) identified for the first time an aromatic ring cleavage product, β,γ-cyclic carbonate of arylglycerol, of a β-O-4 lignin substructure model dimer by ligninolytic cultures of a white-rot basidiomycete, Phanerochaete chrysosporium, and we further found that the ring cleavage was catalyzed by lignin peroxidase (ligninase)(9-11).

The purpose of the present paper is to describe mechanisms of side chain cleavages and aromatic ring cleavages of lignin substructure model compounds and lignin (DHP) by the lignin peroxidase of P. chrysosporium, and by the laccase of C. versicolor, and also to discuss potential applications of lignin degradative systems for the biotechnical processing of lignocellulosic materials.

I. SIDE CHAIN CLEAVAGE OF β-O-4 LIGNIN SUBSTRUCTURE MODEL COMPOUND

We (12) synthesized several oligolignols as substrates for Fusarium solani M-13-1 which was isolated from soil (13), and for Phanerochaete chrysosporium and Coriolus versicolor, both typical lignin-degrading basidiomycetes. These lignin substructure model compounds were used as experimental substrates for cultures of the fungi and their enzymes. The degradation products were isolated and identified by NMR and GC-MS to elucidate degradative pathways and mechanisms of enzyme reactions involved. Since degradative pathways for major lignin substructure model compounds by the fungi and their enzymes were reviewed previously (2,6), further investigations on mechanisms of enzymatic degradation of arylglycerol-β-aryl ether (β-O-4) by lignin peroxidase and laccase are only mentioned in this paper as examples of side chain cleavage reactions of lignin substructure model compounds. We used models for the β-O-4 lignin substructure because it is the most frequent interphenylpropane linkage (40-60%) in lignin ; β-O-4 dimers have also been used as important model compounds to elucidate mechanisms of major chemical degradation reactions of lignin such as acidolysis, hydrogenolysis, and delignification in sulfite and Kraft cookings of wood chips, and pulp bleaching.

Our investigation (14) showed for the first time that the γ-benzyl ether of 4-ethoxy-3-methoxyphenylglycerol-β-vanillin ether [1], as a nonphenolic β-O-4 lignin substructure model compound, is converted to the γ-benzyl ether of 4-ethoxy-3-methoxyphenylglycerol [2] by O-C4 bond

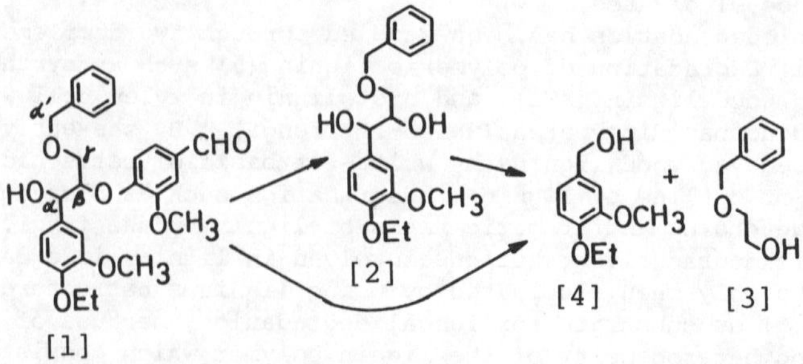

Figure 1. Degradation pathways of a (β-O-4)-(α-O-γ) trimer, the γ-benzyl ether of a β-O-4 lignin substructure model, by P. chrysosporium.

cleavage, to benzyloxyethanol [3] as Cβ-Cγ fragment, and to 4-ethoxy-3-methoxybenzyl alcohol [4] by Cα-Cβ cleavages in the   propyl side chain by ligninolytic cultures of P. chrysosporium (Fig. 1).

Meanwhile, lignin peroxidase (ligninase) which catalyzes Cα-Cβ cleavage of diarylpropane-1,3-diol (β-1) lignin substructure model compounds, was isolated from the culture filtrate of P. chrysosporium by Tien and Kirk (15), and Glenn et al., (16). They found that the enzyme also catalyzes Cα-Cβ clervage of β-O-4 lignin substructure model compounds to give benzaldehydes as C6-Cα fragment, and arylglycerol by O-C4 cleavage, and that $^{18}O_2$ is not incorporated into the benzaldehydes formed as degradation products.

Hence we (17) prepared α,β-dideuterated 4-ethoxy-3-methoxyphenylglyce-rol-β-guaiacyl-γ-benzyl diether [5] as substrate for lignin peroxidase to elucidate the mechanisms for Cα-Cβ cleavage, including formation of the Cβ-Cγ fragment as a degradation product, and O-C4 cleavage. Our experiments showed that the substrate is cleaved to give 4-ethoxy-3-methoxybenzaldehyde [6] and benzyloxyacetaldehyde [7] by Cα-Cβ cleavage, and γ-benzyl ether of 4-ethoxy-3-methoxyphenylglycerol [2] by O-C4 cleavage in agreement with our previous culture experiment. In addition, mass spectrometric analyses showed that deuterium at Cα and Cβ of the γ-benzyl ether of 4-ethoxy-3-methoxyphenylglycerol, of 4-ethoxy-3-methoxybenzaldehyde and of benzyloxy-acetaldehyde were almost quantitatively retained after the Cα-Cβ and O-C4 bond cleavages. These results clearly indicated that hydrogen abstraction was not involved in either the Cα-Cβ or O-C4 cleavages of β-O-4 model compounds. Kersten et al. (18) established that lignin peroxidase catalyzes the one-electron oxidation of aromatic ring of nonphenolic compounds to give cation radicals which undergo spontaneous degradative transformations. Thus, side chain cleavage of β-O-4 model compound by lignin peroxidase can be illustrated as shown in Fig. 2: The Cα-Cβ linkage of the propyl side chain is homolytically cleaved via an aryl cation of the substrate to give benzyl cation and guaiacoxyethyl radical which is attacked by dioxygen to form an unstable hemiketal. The O-Cβ linkage of the hemiketal is subsequently cleaved to give vanillin and benzyloxyacetaldehyde.

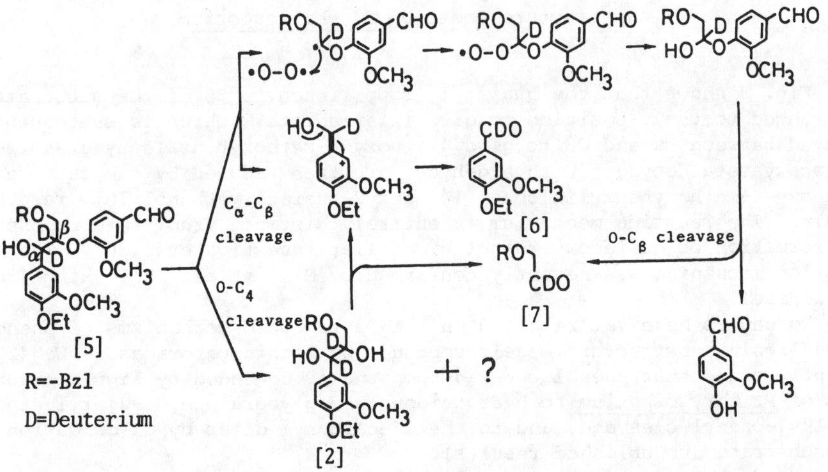

Figure 2. Degradation of a deuterated arylglycerol-β-aryl ether lignin substructure model by lignin peroxidase.

In further investigation we (19) identified an alternative Cα-Cβ cleavage reaction of β-O-4 model compound to give 2-guaiacoxyethanol [8], which was previously identified as a degradation product of β-O-4 model compound by Enoki et al. (20), and benzaldehyde, by ligninolytic cultures of P. chrysosporium. In this experiment we prepared 4-ethoxy-3-methoxy-phenylglycerol-β-^{18}O-guaiacyl ether as substrate for the cultures (19). GC-MS analyses of the isolated degradation products showed that [18]O of the ethereal oxygen of the substrate was not retained in the 2-guaiacoxy-ethanol product. To elucidate the formation mechanism of 2-guaiacoxy-ethanol, 4-ethoxy-3-methoxyphenylglycerol (γ-^{13}C)-β-guaiacyl ether was prepared as substrate for the cultures. Isolated 2-guaiacoxyethanol derived from the substrate was shown by mass spectrometry to be labeled with ^{13}C at the 2-position but not the 1-position. Thus, we proposed the reaction mechanism on the formation of guaiacoxyethanol shown in Fig. 3.

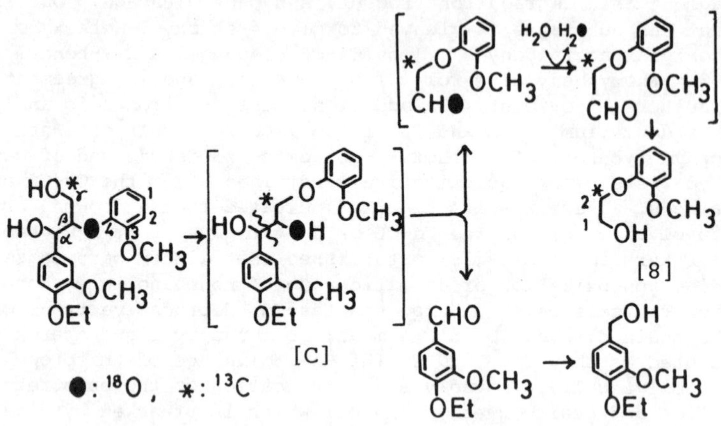

Figure 3. Mechanism of guaiacoxyethanol formation from a β-O-4 lignin substructure model by P. chrysosporium.

Fig. 3 shows that the guaiacyl group linked to Cβ of the substrate is rearranged to the γ-position to give intermediate C which is subsequently cleaved between Çα and Cβ to give 4-ethoxy-3-methoxybenzaldehyde and 2-guaiacoxyacetaldehyde. Both aldehydes are then reduced by the cultures to 4-ethoxy-3-methoxybenzyl alcohol [4] and 2-guaiacoxyethanol [8], respectively. The reaction mechanism is entirely distinct from that proposed for the formation of guaiacoxyethanol by earlier investigaters (20). The reaction mechanism was recently confirmed by Miki et al. (21) using lignin peroxidase.

No papers have yet reported on the degradative mechanisms of phenolic β-O-4 lignin substructure model compounds by lignin peroxidase. We (22) recently found that phenolic β-O-4 compound is degraded by lignin peroxi-dase of P. chrysosporium to hydroquinone and glyceraldehyde-2-aryl ether by alkyl-phenyl cleavage, and to the Cα-carbonyl dimer by Cα-oxidation of the substrate (unpublished results).

For degradation of phenolic β-O-4 compound by laccase of C. versicolor Wariishi et al. (23) recently reported that syringylglycerol-β-guaiacyl ether is cleaved between Cα and Cβ to give syringaldehyde and guaiacoxy-ethanol, 2,6-dimethoxybenzoquinone and glyceric acid-2-guaiacyl ether by alkyl-phenyl cleavage, and guaiacol by O-Cβ cleavage.

However, our investigation (24) on degradation of syringylglycerol-β-guaiacyl ether [9] by laccase of C. versicolor showed that the substrate is mainly converted to the α-carbonyl dimer [10], 2,6-dimethoxyhydroquinone [11], and glyceraldehyde-2-guaiacyl ether [12] by alkyl-phenyl cleavage, and to guaiacol [13] by O-Cβ cleavage. We could not find syringaldehyde and guaiacoxyethanol as direct Cα-Cβ cleavage products of the substrate. Our subsequent investigation to identify the pathway to give guaiacol showed that the α-carbonyl dimer [10] used as substrate is cleaved between Cα and Cβ to give syringic acid [14] and guaiacol as shown in Fig. 4.

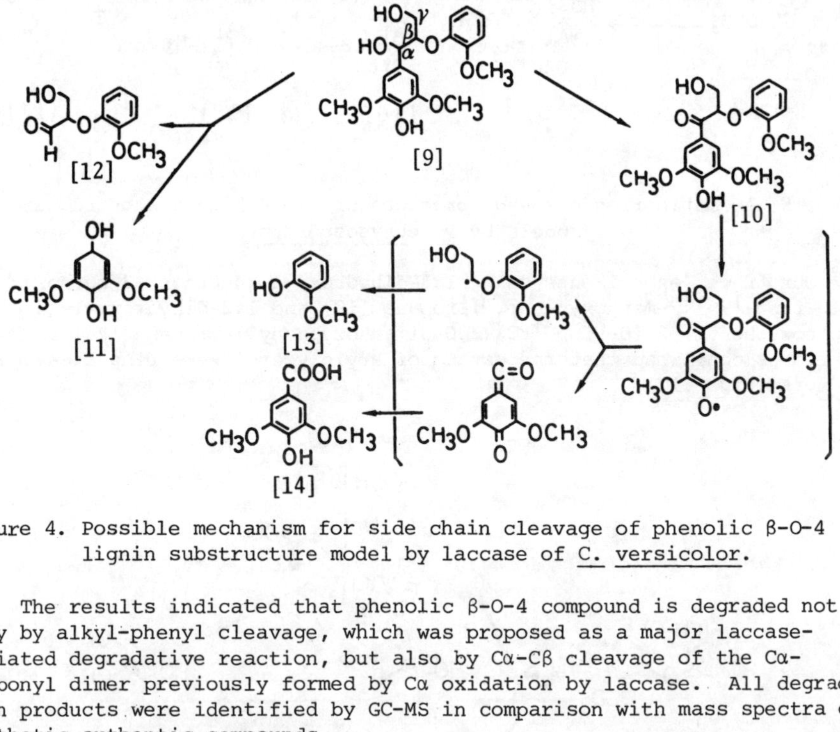

Figure 4. Possible mechanism for side chain cleavage of phenolic β-O-4 lignin substructure model by laccase of C. versicolor.

The results indicated that phenolic β-O-4 compound is degraded not only by alkyl-phenyl cleavage, which was proposed as a major laccase-mediated degradative reaction, but also by Cα-Cβ cleavage of the Cα-carbonyl dimer previously formed by Cα oxidation by laccase. All degradation products were identified by GC-MS in comparison with mass spectra of synthetic authentic compounds.

Recent enzymatic studies (18, 25) showed that lignin peroxidase catalyzes one electron oxidation of both phenolic and nonphenolic lignin substructure compounds to give the corresponding phenoxy radicals and aryl cation radicals. Aryl cation radicals are attacked by nucleophiles such as water or the hydroxy groups of the substrate to produce C-radical intermediates. These C-radical intermediates and phenoxy radicals are either attacked by O_2 radical to give various degradation products, or repolymerized by radical-radical couplings. Laccase only catalyzes oxidation of phenolic substrates to give phenoxy radicals; nonphenolic compounds are not oxidized by laccase.

Our investigations on the side chain cleavage of phenolic β-O-4 lignin substructure model compounds with lignin peroxidase and laccase suggested that the same chemical principle, phenoxy radical as intermediate, is involved in the degradation of phenolic lignin substructure model compounds by both enzymes.

II. AROMATIC RING CLEAVAGE OF NONPHENOLIC β-O-4 MODEL COMPOUNDS AND VERATRYL ALCOHOL BY LIGNIN PEROXIDASE

We (7) identified for the first time an aromatic ring cleavage product of a β-O-4 lignin substructure model compound by ligninolytic cultures of P. chrysosporium: the β,γ-cyclic carbonate [15]. Subsequently, several esters of arylglycerol were identified (8) as products of aromatic ring cleavage of β-O-4 model dimers by the fungus: α,β-cyclic carbonate [16], γ-formate [17], and methyl oxalate [18] (Fig. 5).

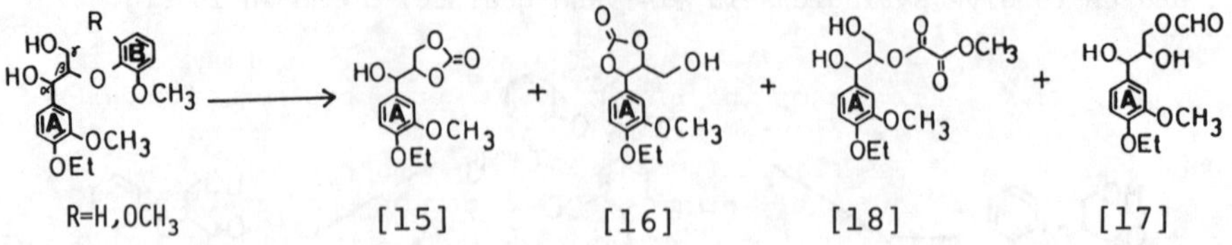

Figure 5. Aromatic ring cleavage products of β-O-4 lignin substructure models by P. chrysosporium.

Our tracer experiments using 1,3-dihydroxy-1-(4-ethoxy-3-methoxyphenyl)-2-[U-ring-^{13}C](2-methoxyphenoxy)propane [19] and 1,3-dihydroxy-1-(4-ethoxy-3-methoxyphenyl)-2-[U-ring-^{13}C](2,6-dimethoxyphenyl)-propane [20] as substrates confirmed that the esters of arylglycerol were ring cleavage products (Fig. 6).

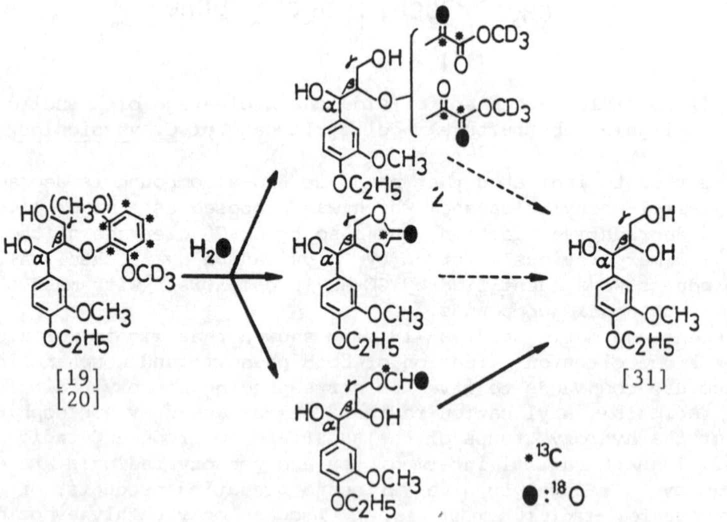

Figure 6. Formation of aromatic ring cleavage products from arylglycerol-β-aryl-U-^{13}C,OCD$_3$ ethers by lignin peroxidase of P. chrysosporium.

We further found that the formation of these aromatic ring cleavage products are not specific to P. chrysosporium but also mediated by other white-rot basidiomycetes, C. versicolor (26) and C. hirsutus (27).

The aromatic ring cleavage products formed by ligninolytic cultures of the fungi were also identified as lignin peroxidase degradation products of β-O-4 model compounds: the cyclic carbonates, the formate, the methyl oxalate, and a novel product, the muconate of arylglycerol [21] which was identified very recently. All the products were confirmed to be aromatic ring cleavage products by mass spectrometry using ring-^{13}C labeled substrates. Subsequently, Miki et al. (28) reported the formation of aromatic ring cleavage products similar to our products in degradation of a β-O-4 substructure model compound by the enzyme.

The muconate [21], which is an immediate aromatic ring cleavage product, retains all six carbon atoms of the B-ring of the substrate. This product therefore seemed to be suitable to examine the mechanism for ring cleavage. GC-MS analyses of the ring cleavage products formed under $H_2^{18}O$ or $^{18}O_2$ showed that one of the carbonyl oxygen atoms of muconate, and of the methyl oxalate, is derived from H_2O and the other from O_2.

Mass spectrometric analyses of the degradation products of 1,3-dihydroxy-1-(4-ethoxy-3-methoxyphenyl)-2-(2-[OC^2H_3]-methoxyphenoxy)propane by the enzyme showed that the methyl group of the methyl ester of the oxalate is derived from the methoxyl group of the B-ring of the substrate (10)(Fig. 6). Thus, the result indicated that demethylation -- which was previously postulated to precede ring cleavage by the fungus (30) is not prerequiste, and that ring cleavage by the enzyme is entirely different from ring cleavage catalyzed by conventional dioxygenases (31). Mass spectrometric analyses also showed that the carbonyl oxygen atoms of the cyclic carbonates and the formate of arylglycerol are derived from H_2O (Fig. 6).

Based on these tracer experiments with ^{13}C, ^2H, and ^{18}O we (29) proposed a general mechanism for aromatic ring cleavage of β-O-4 lignin

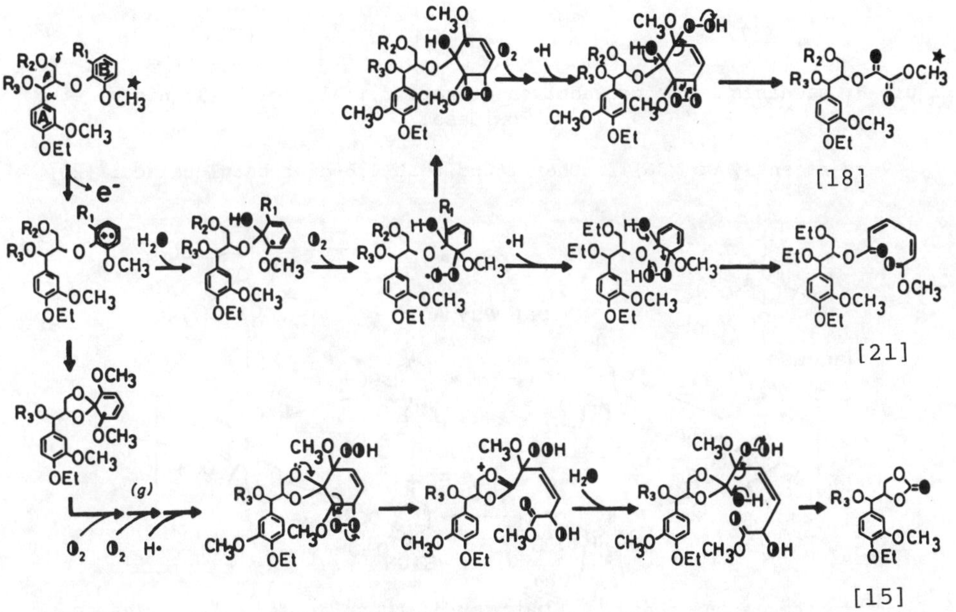

Figure 7. Mechanisms for aromatic ring cleavage of β-O-4 lignin substructure models by lignin peroxidase.

substructure model compounds by lignin peroxidase. One electron oxidation of the B-ring forms the corresponding cation radicals, which are attacked nucleophilically: the remaining radical is attacked by dioxygen (or by a radical species derived from dioxygen)(Fig. 7).

Because these ring cleavage products of β-O-4 model compounds were also formed by C. versicolor and C. hirsutus, it is likely that the aromatic ring cleavages are also mediated by enzymes similar to lignin peroxidase of P. chrysosporium. It was recently shown (32) that C. versicolor produces an enzyme similar to the P. chrysosporium lignin peroxidase.

Leisola et al. (33) reported aromatic ring cleavage of veratryl alcohol [22] by lignin peroxidase of P. chrysosporium, and tentatively identified cis and trans-γ-lactones [23, 24] as ring cleavage products. We (34, 35) confirmed these two γ-lactones as products, and additionally identified a δ-lactone [25]. By tracer experiments with $^{18}O_2$ and $H_2^{18}O$ we found that one oxygen atom each from H_2O and one from O_2 are specifically incorporated into the cleavage products at the original C3 and C4 positions, respectively, in good accordance with the mechanism for ring cleavage of β-O-4 compounds (Fig. 8).

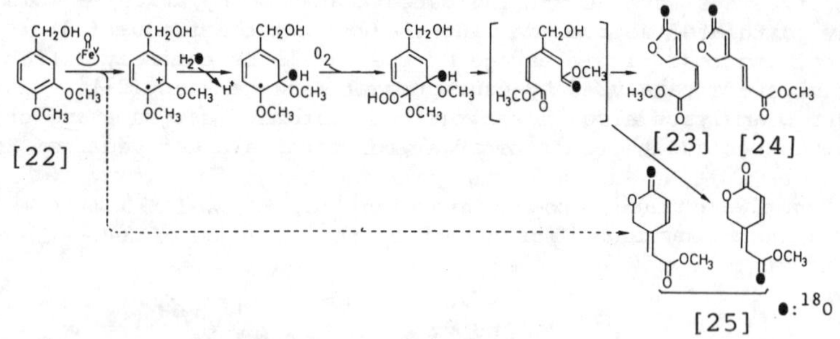

Figure 8. Mechanisms for degradation of veratryl alcohol by lignin peroxidase.

Very recently we (36) further found that 4,6-di-t-butylguaiacol [26]

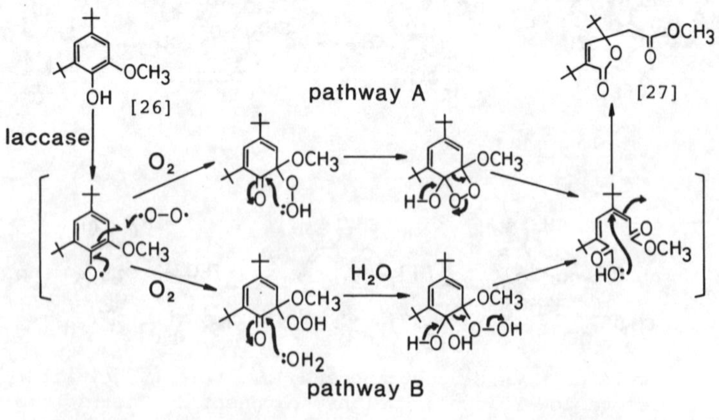

Figure 9. Mechanism for degradation of 4,6-di-t-butylguaiacol by laccase.

is converted by laccase of C. versicolor to a ring cleavage product, the
muconolactone derivative [27], which was previously identified by Gierer
and Imsgard (37) as a product in alkaline-oxygen oxidation of the same
substrate. Our experiment showed that ^{18}O from $^{18}O_2$ but not from $H_2^{18}O$ is
incorporated into [27]. We therefore proposed the following mechanism
(pathway A) for ring cleavage of 4,6-di-t-butylguaiacol by laccase (Fig. 9).

All these studies on the cleavage of side chains and aromatic ring of
lignin model compounds indicated that both lignin peroxidase and laccase,
which catalyze one electron oxidation of either phenolic (both laccase and
lignin peroxidase) or nonphenolic (lignin peroxidase) compounds, are
involved in the initial degradation of lignin substructure model compounds.

III. AROMATIC RING CLEAVAGE OF (SYNTHETIC LIGNIN) BY LIGNIN PEROXIDASE

Tien and Kirk (15) reported degradation of methylated spruce lignin by
lignin peroxidase. They showed formation of low molecular weight degradation fractions by gel filtration with Sephadex LH-20, and detected Cα-Cβ
cleavage products by an isotope trapping method. On the other hand,
Haemmerli et al. (38), and Odier et al. (39) found that polymerization of
the lignin occurred when non-alkylated lignin, with free phenolic hydroxyl
groups, was treated with lignin peroxidase.

These studies indicated that lignin degradation initiated via formation of aryl cation radicals and phenoxy radicals involves intrinsically
repolymerization reactions as well as degradation reactions.

Our previous studies (10) established that aromatic rings of nonphenolic β-O-4 model compounds are cleaved to give, inter alia, a cis,cis-
muconate as an initial ring cleavage product by lignin peroxidase.
However, we still do not know whether or not lignin peroxidase catalyzes
cleavage of aromatic rings, and of O-C4 bonds in the lignin polymer. To
examine this question we(40)synthesized a co-polymer [28] of (β-O-4)-(β-β)
lignin substructure trimer [29] and coniferyl alcohol [30], which is
referred to as (β-O-4)-(β-β)-DHP in this paper (Fig. 10). The (β-O-4)-

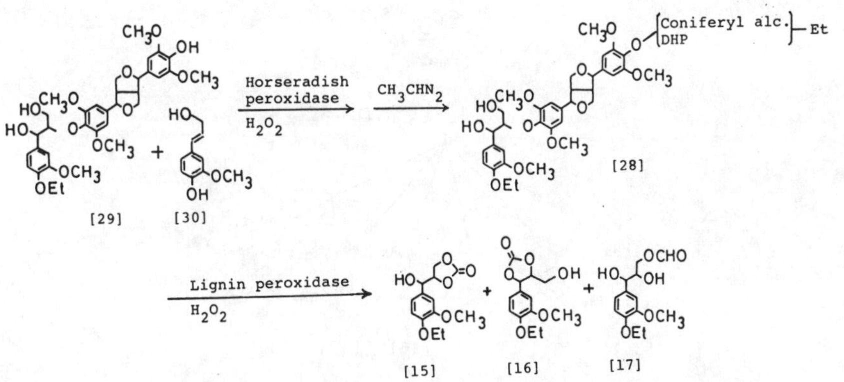

Figure 10. Aromatic ring cleavage of DHP prepared from a mixture of 4-
ethoxy-3-methoxyphenylglycerol-β-syringaresinol ether and
coniferyl alcohol by lignin peroxidase

(β-β)-DHP was ethylated with diazoethane and submitted to gel filtration
(LH-20/DMF) to remove low molecular weight components. The high molecular
weight fraction of the ethylated (β-O-4)-(β-β)-DHP thus obtained was
degraded by lignin peroxidase. The degradation products were extracted

with ethyl acetate, acetylated and partially purified by TLC. GC-MS
analysis of the purified fraction revealed the cyclic carbonates [15] and
[16], formate [17], the arylglycerol [31] and α-carbonylarylglycerol [32]
as products.

The compounds were previously identified as aromatic ring cleavage
products and β-O-4 bond cleavage products of β-O-4 lignin substructure
model compounds by the enzyme. The results indicate that lignin peroxidase
cleaves aromatic rings and O-C4 bonds of synthetic lignin (DHP), in
agreement with our experimental results for β-O-4 dimers.

It is therefore concluded that most of the degradative reactions of
lignin polymers suggested by analyses of decayed lignin by white-rot fungi
(5) were confirmed to be catalyzed by lignin peroxidase and laccase.

IV. BIOMIMETIC APPROACH TO LIGNIN DEGRADATION

Recently biomimetic systems with synthetic porphyrin catalysts mimicking
lignin peroxidase are receiving great attention. The systems seem to be
economically feasible and ideal for practical use for reduction of energy
consumption and environmental pollutions.

We (41) found for the first time that a metalloporphyrin with the
oxidant t-butylhydroperoxide or iodosylbenzene catalyzes oxidative C-C
bond cleavage reactions of the 1,2-bis(4-ethoxy-3-methoxyphenyl)propane-
1,3-diol (β-1) lignin substructure model compound [33] to give 4-O-
ethylvanillin [34], α-hydroxy-4-ethoxy-3-methoxyacetophenone [35], 4-O-
ethylvanillic acid [36], 4-ethoxy-3-methoxyphenylglycol [37], 4-ethoxy-3-
methoxy-α-(4-ethoxy-3-methoxyphenyl)-β-hydroxypropiophenone [38] and
formaldehyde. Our studies (42) further showed that other metaloporphyrins
with H_2O_2 also catalyze oxidative degradation of β-O-4 and β-5 model
compounds as well as β-1 model compounds by Cα-Cβ cleavage and O-C4
cleavage, and aromatic ring cleavage of veratryl alcohol via aryl cation
radicals by single electron oxidation.

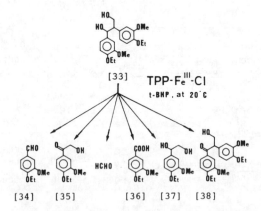

Figure 11. Oxidative degradation of β-1 lignin model compound by catalysis
of TPP(Fe)Cl in the presence of t-BHP.

Recently Dolphin et al. (43) also showed that a series of water
soluble iron-and other metalloporphyrins are good mimics of lignin
peroxidase, and degrade β-O-4 compound and veratryl alcohol oxidatively.
While, Paszczynski et al. (44) reported very recently on delignification
of wood chips and pulps by natural and synthetic porphyrins as models

of fungal decay. They concluded that the heme-t-butyl hydroperoxide system which was first used by us mimics the decay of lignified tissues by white-rot fungi. Thus, the use of metalloporphyrins appears to be suitable for catalytic degradation of the lignin polymer, although it is still premature to use the presently available metalloporphyrin catalysts for industrial pulping processes.

This paper was prepared based on our recent work on lignin biodegradation in the Research Section of Lignin Chemistry, Wood Research Institute. The author is indebted to Dr. M. Shimada, Dr. T. Umezawa, Messrs. S. Kawai, S. Yokota, and T. Hattori in this section for their cooperation.

REFERENCES

1. Kirk, T.K. and Farrell, R.L., Ann. Rev. Microbiol., 1987, 41, 465-505.
2. Higuchi, T., Wood Research, 1986, 73, 58-81.
3. Buswell, J.A. and Odier, E., CRC Critical Rev. Biotechnol., 1987, 6, 1-60.
4. Lignin Enzymic and Microbial Degradation, Odier, E. ed., INRA, Paris, 1987.
5. Chen, C.-L. and Chang, H.-m., In Biosynthesis and Biodegradation of Wood Components, Higuchi, T. ed., Academic Press, Orlando, FL, 1985, pp. 535- 56.
6. Higuchi, T., In Biosynthesis and Biodegradation of Wood Components, Higuchi, T., ed., Academic Press, Orlando, FL, 1985, pp. 557-78.
7. Umezawa, T. and Higuchi, T., FEBS Lett., 1985, 182, 257-9.
8. Umezawa, T., Kawai, S., Yokota, S. and Higuchi, T., Wood Research, 1986, 73, 8-17.
9. Umezawa, T., Shimada, M., Higuchi, T. and Kusai, K., FEBS Lett., 1986, 205, 287-92.
10. Umezawa, T. and Higuchi, T., FEBS Lett., 1986, 205, 293-98.
11. Umezawa, T. and Higuchi, T., Agric. Biol. Chem., 1987, 51, 2281-84.
12. Higuchi, T. and Nakatsubo, F., Kemia-Kemi, 1980, 9, 481-88.
13. Higuchi, T., In Lignin Biodegradation: Microbiology, Chemistry, and Potential Applications, Vol. I., Kirk, T.K., Higuchi, T. and Chang, H.-m. ed., CRC Press, Boca Raton, FL, 1980, pp. 171-92.
14. Umezawa, T., Nakatsubo, F. and Higuchi, T., Agric. Biol. Chem., 1983, 47, 2677-81.
15. Tien, M. and Kirk, T.K., Science, 1983, 221, 661-63.
16. Glenn, J.K., Morgan, M.A., Mayfield, M.B., Kuwahara, M. and Gold, M.H., Biochem. Biophys. Res. Commun., 1983, 114, 1077-83.
17. Habe, T., Shimada, M., Umezawa, T. and Higuchi, T., Agric. Biol. Chem., 1985, 49, 3501-10.
18. Kersten, P.J., Tien, M., Kalyanarama, B. and Kirk, T.K., J. Biol. Chem., 1985, 260, 2609-12.
19. Umezawa, T. and Higuchi, T., FEBS Lett., 1985, 192, 147-50.
20. Enoki, A., Goldsby, G.P. and Gold, M.H., Arch. Microbiol., 1980, 125, 227-32.
21. Miki, K., Renganathan, V. and Gold, M.H., Proc. 3rd Intern. Conf. Biotechnol. Pulp Paper Ind., Stockholm, 1986, pp. 13-6.
22. Yokota, S., unpublished.
23. Wariishi, H., Morohoshi, N. and Haraguchi, T., Mokuzai Gakkaishi, 1987, 33, 892-98.
24. Kawai, S., Umezawa, T. and Higuchi, T., unpublished.
25. Higuchi, T., Proc. Intern. Symp. Biosynthesis and Biodegradation of Plant Cell Wall Polymer, Toronto, June 5-10, 1988, in press.
26. Kawai, S., Umezawa, T. and Higuchi, T., App. Environ. Microbiol., 1985,

50, 1505-8.

27. Yoshihara, K., Umezawa, T., Higuchi, T. and Nishiyama, M., Agric. Biol. Chem., in press.
28. Miki, K., Renganathan, V., Mayfield, M.B. and Gold, M.H., FEBS Lett., 1987, 210, 199-203.
29. Umezawa, T. and Higuchi, T., FEBS Lett., 1987, 218, 255-60.
30. Chen, C.-L., Chang, H.-m. and Kirk, T.K., J. Wood Chem., Technol., 1983, 3, 35-57.
31. Cain, R.B., In Lignin Biodegradation: Microbiology, Chemistry and Potential Applications, Kirk, T.K., Higuchi, T. and Chang, H.-m. ed., CRC Press, Baca Raton, FL, Vol. I., 1980, pp. 21-60.
32. Dodson, P.J., Evans, C.S., Harvey, P.J. and Palmer, J.M., FEBS Microbiol. Lett., 1987, 42, 17-22.
33. Leisola, M.S.A., Schmidt, B., Thanei-Wyss, U. and Fiechter, A., FEBS Lett., 1985, 187, 267-70.
34. Shimada, M., Hattori, T., Umezawa, T., Higuchi, T. and Uzura, K., FEBS Lett., 1987, 221, 327-31.
35. Hattori, T., Shimada, M., Umezawa, T., Higuchi, T., Leisola, M.S.A. and Fiechter, A., Agric. Biol. Chem., 1988, 52, 879-80.
36. Kawai, S., Umezawa, T., Shimada, M. and Higuchi, T., FEBS Lett., submitted.
37. Gierer, J. and Imsgard, F., Acta Chem. Scand., 1977, 31, 546-60.
38. Haemmerili, S.D., Leisola, M.S.A. and Fiechter, A., FEMS Microbiol. Lett., 1986, 35, 33-6.
39. Odier, E., Monzuch, M., Kalyanaraman, B. and Kirk, T.K., In Lignin Enzymic and Microbial Degradation, Odier, E. ed., INRA, Paris, 1987, 131-6.
40. Umezawa, T. and Higuchi, T., FEBS Lett., submitted.
41. Shimada, M., Habe, T., Umezawa, T., Higuchi, T. and Okamoto, T., Biochem. Biophys. Res. Commun., 1984, 122, 1247-52.
42. Shimada, M., Habe, T., Higuchi, T., Okamoto, T. and Panijpan, B., Holzforschung, 1987, 41, 277-85.
43. Dolphin, D., Nakano, T., Maione, T.E., Kirk, T.K. and Farrell, R., In Lignin Enzymic and Microbial Degradation, Odier, E. ed., INRA, Paris, 1987, pp. 157-62.
44. Paszczynski, A., Crawford, R.L. and Blanchette, R.A., App. Environ. Microbiol., 1988, 54, 62-8.

DISCUSSION

Q: A. Blažej (Slovak Tech. Univ.)
What do you think on presence of veratryl alcohol like radical cation as a mediator for lignin degradation ? Do you probably know this mechanism suggested by Harvey, Schoemaker and Palmer ?

A: I know that Dr. Harvey and coworkers proposed that veratryl alcohol plays as a mediator to degrade lignin. However, I doubt that veratryl alcohol acts as a radical mediator in lignin biodegradation. Veratryl alcohol is mostly oxidized to veratryl aldehyde, and in fact Dr. Hatakka in Finland found that degradation of β-O-4 lignin substructure compound by lignin peroxidase was not promoted in the presence of veratryl alcohol.

BIOCONVERSION OF WOOD AND CELLULOSIC MATERIALS

JOHN F. KENNEDY AND MARION PATERSON
Research Laboratory for the Chemistry of
Bioactive Carbohydrates and Proteins,
Department of Chemistry, The University of Birmingham,
Birmingham B15 2TT, U.K.

SUMMARY

The activities of forestry and agriculture industries generate very large quantities of waste that are currently not used for food, fuel or construction purposes. Much of this material is burned or allowed to decay naturally, often producing environmental problems such as contamination of waterways and air pollution.

This paper outlines what is currently or might in the future be economically produced from these lignocellulosic materials by using bioconversion processes.

INTRODUCTION

Wood, as a renewable resource, represents an important source of organic chemicals and materials for the future. The three major components of wood are cellulose, hemicellulose and lignin. Various routes have been devised by which to obtain organic chemicals from these components [1]. Nevertheless an immediate and pressing need is the identification of feasible and more efficient conversion processes to assist wood-dependent nations in making better use of their wood supplies.

In terms of potential technical applications there are two categories cellulosic materials. The long fibres of many trees, especially softwood species, are most valuable for the pulp and paper industry, whereas agricultural residues, other non-wood materials or very fast growing trees are less suitable for this purpose. In the former case, biotechnical applications are closely associated with the chemical processing of wood. In these processes the feedstock for fermentation processes is derived entirely from hemicellulose, whereas both the cellulose and the hemicellulose of agricultural residues are potential feedstocks for fermentation. Concerning the utilization of lignin, efforts have been made to use it in an economical way for purposes other than energy generation.

In general, woody material and agricultural residues are not in a suitable state for immediate bioprocessing. Accessibility of enzyme systems to the polysaccharides is restricted by the presence of the lignin matrix which acts as a barrier. Furthermore, the crystallinity of the cellulose which is the major biomass component also serves to retard the rate of production of fermentable sugars. Preliminary preparation or pretreatment of the biomass to overcome these impediments is usually necessary.

This paper outlines various bioconversion processes by which lignocellulosic materials can be converted into products of present and potential importance.

WOOD

Wood can be obtained from three basic sources. These are mill residues, wood pellets, and standing timber.

The chemical composition of wood varies from one species to another. About 40–50% of the dry substances in most woods are cellulose, located mainly in the secondary cell wall [2]. Cellulose is organised into laminar crystallites which are bundled into the microfibrils. Each microfibril contains regions of amorphous cellulose interspersed and interwined with hemicellulose [3]. The $(1 \longrightarrow 4)-\beta$-D-glucose chains which make up the cellulose polymer can be arranged in both parallel and antiparallel crystallites. However, it has now been accepted that all cellulose chains in native microfibrils are orientated in the same direction.

Hemicelluloses are the principal non-cellulosic fraction of wood polysaccharides and its content in wood is usually between 20–30% on a dry basis [2]. The role of this component is to provide a linkage between lignin and cellulose. In its natural state, it exists in an amorphous form and can be divided into three groups, namely, xylans, mannans and galactans.

Lignin is essentially a three dimensional phenylpropane polymer with phenylpropane units held together by ether and carbon-carbon bonds. The amount of lignin constitutes 20–35% of the wood structure [4]. In wood, the lignin network is concentrated between the outer layers of fibres and gives structural rigidity by stiffening and holding the fibres of polysaccharides together. The lignin from grasses, softwoods, and hardwoods differs somewhat in composition, mainly in methoxyl substitution, and the degree of linkage between phenyl groups.

SOURCES OF CELLULOSIC MATERIAL

Table 1 lists the potential sources and diverse nature of various cellulosic materials available for bioconversion to fuel, food and chemical feedstocks. Such biomass may be residues resulting from other sources or may be directly harvested. The raw biomass which is likely to be quite variable in composition, texture, and moisture content must be converted to a relatively homogeneous feedstock to be used in different fermentation processes.

TABLE 1
Potential sources of cellulosic materials

Item	Description
Crop residue	Sugarcane waste, weeds, corn and other stuble, straw, wasted fodder
Forest residue	Twigs, bark, sawdust, branches, undergrowth
By-products/wastes from agricultural related industries	Bagasse, rice bran, seeds, cotton dust (textile industry)
Solid wastes/urban	Paper wastes, vegetable wastes

BIOCONVERSON OF CELLULOSE

The hydrolysis of cellulose or cellulosic materials to glucose or to other monosaccharides, either by acids or enzymes, is the most common method of converting them to useful products [5,6].

The feasibility of hydrolysing wood and cellulosic materials by enzymes has been studied intensively in recent years. Also, the heterogeneity of the biomass is a big problem. Compared with acid hydrolysis, enzymatic degradation proceeds in non-corrosive conditions and the formation of harmful by-products is reduced.

The group of hydrolytic and other enzymes which perform the biodegradation of cellulose are known as cellulases.

Cellulases are produced by many microorganisms such as bacteria, actinomycetes and fungi, by higher plants and also by some invertebrate animals [7]. The most widely studied cellulases are of fungal origin, particularly the white, brown and softwood rotting fungi. Trichoderma species are the most potent cellulase producers of this group.

The commercial potential for the rapid enzymatic degradation of cellulose is enormous. Saccharification of cellulose is a prerequisite to develop processes for the conversion of cellulose into products such as ethanol, methanol, methane etc. The most useful product of cellulose hydrolysis would be glucose, but most cellulases produce significant quantities of cellobiose, and the composition of syrups depends upon the activity of cellobiose-hydrolysing and oxidising enzymes.

The raw materials used in commercial methane generation have been traditionally classified as waste materials, which include crop residues and various urban wastes (paper wastes).

The biogas system consists of anaerobic digestion of biologically degradable cellulosic material. The amount and quality of gas produced depends on the biomass used.

The anaerobic digestion of the waste material is often visualised as a three stage process [8]. The first stage consists of facultative microorganisms attacking the organic matter. Polymers are transformed into soluble fermentable monomers through enzymatic hydrolysis. These monomers become the substrates for the microorganisms in the second stage where soluble organics are converted into organic acids. Soluble organic acids, consisting primarily of acetic acid, form the substrates for the third stage. In this last step, methanogenic bacteria, which are strictly anaerobic in nature, can generate methane by two different routes: one is by fermenting acetic acid to methane and carbon dioxide, whereas the other consists of reducing carbon dioxide to methane via hydrogen gas or formate generated by other bacterial species. Biological growth takes place during all three stages of the fermentation processes.

Considerable effort has gone into various processes to produce glucose and/or ethanol directly or indirectly from lignocelluloses. The routes are: first to pretreat the lignocellulose material to reduce the particle size and then either to delignify and biologically hydrolyse the cellulose to sugars or to hydrolyse the whole lignocellulose chemically and ferment the sugars to ethanol using yeast. The major problems are that the process produces bulk chemicals such as sugars or ethanol at prices which are non-competitive with the corresponding products from agriculture or chemical feedstocks.

n-Propyl alcohol can also be produced in small quantities as a by-product in the distillation of fermented products.

In the fermentation process for n-butyl alcohol production, the carbohydrate mash is acted on by a bacterial culture, namely Clostridium saccharobutyl acetonicum liquefaciens [8].

For the production of single cell protein (SCP) from lignocellulose there are two major bioconversion routes. One is to use the lignocellulose material relatively intact, or with minimal size reduction, and treat it with microorganisms to upgrade it to feed-material [9,10]. The other route to SCP production is to pretreat and/or hydrolyse the lignocellulose material to produce SCP directly from cellulose or from glucose or sugars produced by hydrolysis [10].

BIOCONVERSION OF HEMICELLULOSES

In terms of biotechnology being integrated in the pulp and paper industry, the hemicellulose fraction could be the most important of the three main components of lignocellulose.

In the sulphite process hemicellulose is solubilised and hydrolysed primarily to monomeric sugars, which serve as carbon sources for the fermentation. Unfortunately sulphite processes have been largely replaced by sulphate (Kraft) processes, in which the monosaccharides derived from hemicellulose are decomposed to isosaccharinic acids and other compounds, which result in a non fermentable black syrup being formed. This makes the role of biotechnology extremely difficult.

207

Appreciable quantities of xylan, the main component of plant
hemicelluloses, are currently not utilized as it would be desired.

Xylan and xylooligosaccharides can be converted to xylose by chemical
or enzymic hydrolysis. Subsequently xylose can be converted biologically
to SCP and to a whole range of fuels and chemicals, and can also be used
as a carbon source for production of glucose isomerase [11], (see figure
1). The chemical hydrolysis of xylan accomplished by acids is rapid;
however, xylose is partially degraded to furfural which may hinder
subsequent microbial fermentations. For these reasons, considerable
attention is devoted to development of hydrolytic processes under mild
conditions, which include also the enzymic hydrolysis.

The complex structure of the substrate and, consequently, the
requirement for a special composition of hydrolytic enzymes may represent
a holdup of industrial applications of isolated enzymes in xylan
conversions.

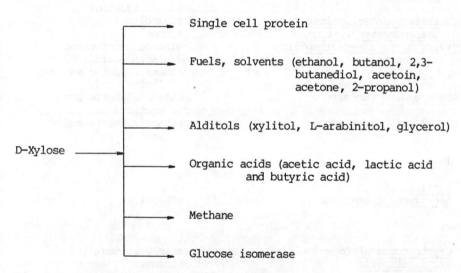

Figure 1.   Products which can be obtained from xylose by microbial
fermentation.

The most promising approach to elaboration of economically feasible
processes of xylan bioconversions is the use of microorganisms capable of
transforming xylan directly to SCP, fuels and chemicals. The
microorganisms which are able to ferment xylose to ethanol are the
Fusarium oxysporum, Aureobasidium pullulans and Candida shehatae [11]. For
direct xylan bioconversions only anaerobic thermophilic bacteria and a few
representatives of fungi and yeasts combine the two metabolic pathways
necessary for direct xylan bioconversions (see Table 2). In general,
yeast is a better choice for fermentation of glucose than other
microorganisms due to its high fermentation rate, ethanol tolerance,
yield, and nonpathogenicity, while bacteria produce undesirable byproducts such as lactic acid and acetic acid along with ethanol. Although
the bacterium Zymomonas mobilis [12] has been reported to produce a high
concentration of ethanol from glucose with little by-product formation, it

has not been clearly determined whether or not this organism can ferment D-xylose or D-xylulose to ethanol.

TABLE 2

Fermentation abilities of xylanolytic microorganisms using xylan as substrate

Microorganism	Product
Bacteria	
Clostridium acetobutylicum	butanol, butyric acid
C. thermocellum	ethanol, acetic and lactic acid
C. thermohydrosulfuricum	ethanol
C. thermosaccharolyticum	ethanol
Thermoanaerobacter ethanolicus	ethanol, acetic and lactic acid
T. brockii	ethanol, acetic and lactic acid
Thermoanaerobium spp.	ethanol, acetic and lactic acid
Thermobacteroides spp.	ethanol, acetic and lactic acid
Fungi	
Monilia spp.	ethanol
Neurospora crassa	ethanol
Yeasts	
Cryptococcus albidus	triglycerides
Pichia stipitis	ethanol

BIOCONVERSION OF LIGNIN

Lignins are available as a natural waste or by product and represent an underused resource. Degradation of lignin by microorganisms is hindered by the unique structure of the polyphenolic structure of lignin. Because of the recalcitrancy of lignin to microbial degradation, lignin and lignin derived compounds are a major source of pollution in wood pulping and processing industries. Knowledge of the physiology of lignin degrading microorganisms and biochemistry of lignin biodegradation may lead to the development of processes for the use of lignin as a renewable resource of chemical feedstocks for industries.

Lignin degradation in nature is the result of the cooperative action between fungi and bacteria. The ability to degrade or modify lignin is restricted to relatively few microorganisms, viz. the bacterial population

of soils [13,14] and waste liquors [15,16], and the wood- and litter-
degrading basidiomycetes [13].

The possibilities of an enzymic approach to the bioconversion of
lignins offers several potential advantages over chemical processing.
These are, greater substrate and reaction specificity, lower energy
requirements, lower pollution generation, higher yields of desired
products and opportunities for transformations not feasible with chemical
reagents [17].

TABLE 3
Applications of waste lignin

Treatment	Product
Alkaline oxidation	Vanillin
	Vanillic acid
	Acetovanillone
	Syringaldehyde
	Organic acid
Alkaline fusion and demethoxylation	Phenol
	Cresol
	Organic acids
	Dimethylsulphide
	Demethylated
	lignin
Hydrogenolysis	Phenol
	Propyl-phenols
	Cyclohexane
	Benzene
	Tar
Pyrolysis	Hydrocarbons
	Phenol
	Aliphatic
	Tar
	Charcoal
Use of polymer	Fuel
	Stabilizers
	Dispersants
	Ion exchangers
	Adhesives

There are two basic approaches to the development of such
bioconversions. The biochemical approach would be to alter the structure
of polymeric lignin microbiologically to enhance its later chemical
degradation to high yields of specific phenolic chemicals. The direct
bioconversion approach would be to use a microorganism specifically as the

catalytic agent for degrading lignin to low-molecular-weight phenolics
[14]. In both approaches, genetic manipulation of promising microbial
strains, to enhance the production of desired products, is likely to be
very important to the successful development of bioconversion of lignin to
chemicals.

Lignins are an ideal choice for agricultural controlled release
formulations [18]. Further, they form a glassy matrix when physically
incorporated into certain pesticides [19]. Lignins can also be esterified
with active agents possessing suitable functional groups or can be crosslinked to form a gel into which the active agent can be incorporated by
swelling, in an aqueous medium and then drying [20]. Lignin can also be
chemically modified to serve as a nutrient source or as a pesticide in its
own right. Low molecular weight phenols produced by degradation of lignin
might have a large potential possibility in the utilization of lignin.
Table 3 shows some industrial applications of lignin.

PROSPECTS FOR BIOCONVERSIONS OF CELLULOSIC MATERIALS

To justify processing low value forestry or agricultural wastes,
significant value must be added to the final product to allow it to
compete with products from other routes, particularly where materials can
be produced by chemical or agricultural routes.

Considerable efforts have been expended in areas such as
microbiology, biochemistry, genetic engineering and biochemical
engineering to improve the efficiency of the bioconversion of
lignocellulosic materials and to overcome some of the barriers inherent in
the structure of the material.

The most recent area of progress has been to clone cellulase or
hemicellulase genes into non-cellulolytic organisms with a view to
improving cellulase production in the producer organisms.

It is expected that, with a better knowledge of the structural and
material properties of the cellulose, hemicellulose and lignin, new areas
will be opened up in future for product development based on bioconversion
of these macromolecules.

REFERENCES

1. Fengel, D. & Wegener, G. Wood, Chemistry. Ultrastructure,
    Reactions. Walter de Gruyter, Berlin, 1984, pp. 526-566.

2. Nakano, J. In: Kennedy, J.F., Phillips, G.O. & Williams, P.A.
    (eds) Wood and Cellulosics, Ellis Horwood Limited, Chicester, 1987,
    pp. 499-512.

3. Jeffries, T.W. In: Kennedy, J.F., Phillips, G.O. & Williams, P.A.
    (eds) Wood and Cellulosics, Ellis Horwood Limited, Chicester, 1987,
    pp. 213-230.

4.   Fan, L.T., Gharpuray, M.M. & Lee, Y.-H. In Aiba, S., et al (eds)
     Cellulose Hydrolysis, vol. 3, Springer-Verlag, Berlin, 1987, pp. 5-20.

5.   Ander, P. & Eriksson, K.E. In: Bull, M.J. (ed) Progress in Industrial
     Microbiology, vol. 14, Elsevier, New York, 1978.

6.   Brown, R.D. & Jurasek, L. (eds) Hydrolysis of Cellulose; Mechanisms
     of Enzymatic and Acid Catalysis, Advances in Chemistry series 181,
     American Chemical Society, Washington D.C., 1979.

7.   Finch, P. & Roberts, J.C. In: Nevell, T.P. & Zeronian, S.H. (eds)
     Cellulose Chemistry, Ellis Horwood Limited, Chichester, 1985, pp.
     312-343.

8.   Cheremisinoff, N.P., Cheremisinoff, P.N. & Ellerbusch In: Powers,
     P.N. (ed) Biomass, Marcel Dekker, Inc., New York, 1980.

9.   Bellamy, W.D. In: Ledward, D.A., Taylor, A.J. & Lawrie, R.A. (eds)
     Upgrading Waste for Feeds and Foods, Butterworths, London, 1983, pp.
     141-152.

10. Rolz, C. (1984). Microbial biomass from renewables. A second review
     of alternatives. Ann. Rep. Ferment. Processes 8: pp. 213-355.

11. Blazej, A. & Biely, P. In: Kennedy, J.F., Phillips, G.O. & Williams,
     P.A. (eds) Wood and Cellulosics, Ellis Horwood Limited, Chichester,
     1987, pp. 275-281.

12. Lee, K.J., Tribe, D.E. & Rogers, P.L. (1979) Biotechnol. Lett., 1,
     421.

13. Crawford, D.L. & Crawford, R.L. (1980) Enzyme Microb. Technol. 2:
     11-22.

14. Crawford, D.L. (1981) Biotechnology and Bioengineering Symp. No. 11,
     275-291.

15. Colberg, P.J. & Young, L.Y. (1982) Can. J. Microbiol. 28: 886-889.

16. Pellinen, J., Vaisanen, E., Salkinoja-Salonen, M., & Brunow, G.
     (1984) Appl. Microbiol. Biotechnol. 20: 77-82.

17. Harvey, P.J., Schoemaker, H.E. & Palmer, J.M. (1985). Ann. Proc.
     Phytochem. Soc. Eur. 26 pp. 249-266.

18. Chanse, A. & Wilkins, R.M. In: Kennedy, J.F., Phillips, G.O. &
     Williams, P.A. (eds) Wood and Cellulosics, Ellis Horwood Limited,
     Chichester, 1987, pp. 385-391.

19. Wilkins, R.M. (1983) Lignins as formulating agents for controlled
     release in agriculture, Brit. Polymer J., 15, 177-78.

20. Dellicolli, H.T. In: Kydonieus, A.F. (ed.) Pine Kraft lignin as a
     pesticide delivery system in Controlled Release Technologies, CRC,
     Boca Raton, Florida, 1980, 225-34.

DISCUSSION

Comment: R. P. Overend (Nat. Res. Council Canada)
    Biomass (lignocellulosic) utilization via biotechnologi-
cal processes invariably involve the use of high water
dilutions ; e.g. most lignocellulosics can only make 5 wt%
slurries before the rheology is limiting. The result is a
large waste treatment burden - the Achilles heel of biotechnology routes - both environmentally and economically.

A: Accepted that large amount of water are involved in the
processes which would be used in industry, but two phenomena
which are being actively researched could counteract this
problem : Firstly, in the last three years attention has been
given to ambient temperature extraction of organic materials
from dilute aqueous solution. Secondly, the current successful studies of use of enzymes in high solution concentrations
of substrate, and in partial solvent systems.

Q: T. Higuchi (Kyoto Univ.)
    How different is red rot fungi from white rot fungi ?

A: It is known that a wide range of microorganisms namely
bacteria, fungi, etc., from different taxonomic groups are
capable of destroying wood. White rots and red rots destroy
both cellulose and lignin in the wood cell walls, whereas
brown rots only attack cellulose (Schurz, J., In: Ghose, T.K.
"Bioconversion of Cellulosic Substances into Energy,Chemicals
and Microbial Protein", Symp. Proc. BERC II T Delhi, New
Dehli, 1978). However, the extent of degradation of wood by
microorganisms depend on their taxonomic identity. They show
a different micromorphology which will determine their capacity of degrading the components of wood (Levy, J.F.,
In: Hartley, B.S., Broda, P.M.A. and Senior P.J. "Technology
in the Utilization of Lignocellulosic Wastes", The Royal
Society, London, 1987).

ROUND TABLE II-2   <Bio-Tech>

"BIOACTIVE (PHARMACOLOGICAL) PROPERTIES"

Chairman: C. Schuerch
           (SUNY College, Enviro. Sci. Forest.)

ANTITUMOR ACTIVITY OF GLUCANS AND THEIR SEMISYNTHETIC DERIVATIVES

Andreas Hensel and Gerhard Franz[*]
University of Regensburg, Faculty of Chemistry and Pharmacy,
Universitätsstraße 31, D-8400 Regensburg (F.R.G.)
[*]Author for correspondence

ABSTRACT

Antitumor effects against allogeneic Sarcoma 180 of several glucans (oat glucan, barley glucan, lichenin, isolichenin) and xyloglucans were investigated. Tumor inhibition rates correlated with the degree of helical proportions in these glucans. Linear glucans such as barley glucan or xyloglucan have no physiological effect. Synthetic modifications of barley glucan enhanced the antitumor activity. Both, long chain substituents and acidic groups increased the antitumor effect: Most active was the combination of both as shown with propoxy-sulfo-glucan derivatives.

INTRODUCTION

The discovery that certain polysaccharides exhibit significant inhibition of tumor growth in the therapy of malignant diseases, led in the last decade to the isolation and screening of a great number of so-called antitumor polysaccharides.

Investigations on the mode of action suggested that activity on NK- cells, cytotoxic macrophages and cytotoxic T-lymphocytes is clearly stimulated. It was further shown that there was no direct cytotoxic effect on tumor cells in vitro (1,2,3).

Several studies of the polysaccharide structures hitherto tested demonstrated the prominent activity of ß-1.3-linked glucans with a certain degree of side chains in position 6 (1). But also completely different polysaccharides like fructans (3), glucomannans (4) or heteropolysacchari-des (5) seemed to be highly active.

The question about essential structural features for polysaccharides with antitumoral properties remained unclear.

In this study it was tried to find out relations between structure and biological effects, by utilizing a series of established non-cellulosic glucans from higher plants. These glucans were isolated, characterized in their structure and tested in a screening system on antitumor activity. By following semisynthetic alterations of these glucans it should be possible to obtain more indications about the necessary structural features of immunmodulating polysaccharides.

MATERIALS AND METHODS

Isolation of oat glucan: Hemicellulose B was extracted from 8 days old Avena sativa seedlings (13). After ion exchange chromatography on DEAE-Sephacel (Pharmacia) the neutral fraction was dialysed and dried. Two times 10% (w/w) of ammonium sulfate was added to a 10% solution of this fraction.

The precipitate obtained at 20% saturation was removed, dissolved in 8 M urea, dialysed and dried. The pure high molecular glucan was separated on Sephacryl-500 (Pharmacia).

Isolichenin was isolated according to (14), arabinoxyloglucan to (7), galactoxyloglucan to (8).

Barley glucan was a commercial product from Biocon Biochemicals, lichenin from Roth GmbH.

Preparation of 6-0-glucosyl-glucans was done according to (9), glucan ethers to (15,16), glucan esters to (17).

Antitumor assay were carried out as described by Kraus et al. (3), statistical evaluations were calculated with the aid of students T-Test (18).

RESULTS

High molecular weight, linear β-1.3/1.4-glucans frequently occur as cell wall components in poaceae plants. One of these glucans could be isolated from 8 days old seedlings of Avena sativa. A similar structured polysaccharide, the so-called barley glucan could be isolated from Hordeum vulgare. Two additional non-cellulosic glucans were obtained from the lichens Cetraria islandica and Cetraria tenuifolia, i.e. the ß-configurated lichenin and the α-glucan isolichenin. The structural characteristics of these glucans are shown in table 1.

For a successful screening of hypothetic antitumor activity, these glucans were tested according to the method of Tabata et al., modified by Kraus (3) on implanted Sarcoma 180, an allogeneic and therefore sensitive tumor system. As shown in table 2, only oat glucan and lichenin exhibited significant inhibition of tumor growth. Neither barley glucan nor isolichenin showed higher retardation.

TABLE 1
DP and structural features of non-cellulosic glucans

	Barley-glucan	Oat-glucan	Lichenin	Isolichenin
molecular weight	630 000	500 000	17 000	69 000
DP	3 889	3 087	105	426
1.3-linkages	32%	34%	33%	34%
1.4-linkages	68%	66%	66%	65%
configuration	β	β	β	a

TABLE 2
Antitumor activity of non-cellulosic glucans
(Sarcoma 180 / CD 1 mice)

substance	dose (mg/kg)	tumor weight (g)	inhibition (%)	regression
control	--	5,66	--	0/15
oat glucan	25	1,24	78[a]	5/10
	5	1,54	73[b]	5/10
control	--	8,20	--	0/9
barley glucan	25	7,80	6	1/6
control	--	5,66	--	0/12
lichenin	25	0,33	94[a]	6/9
isolichenin	25	2,83	50	2/9

a) significant $p < 0,01$
b) significant $p < 0,05$

Looking at the main structured units of the highly active antitumor polysaccharides, β-1.3 glucans with side chains in position 6 are predominant. For this reason it was not worthwhile to test xyloglucans, which are known to have β-1.4-glucan main chain and xylose, galactose residues in the side chains. Since cellulose has been shown not to be active (6), the question could be clarified, if this effect is only a solubility problem or if ß-1.4-linked glucans are ineffective in general. Therefore an extracellular arabinoxyloglucan from suspension cultured Nicotiana tabacum cells (7) and a galactoxyloglucan from Tropaeolum majus (8) was isolated. While the arabinoxyloglucan had a molecular weight of 95 000 d and side chains of 1.2-linked xylose residues with terminal arabinose residues, the galactoxyloglucan showed a molecular weight of 210 000 d and side chains of xylose and terminal galactose residues. Testing of these polymers with the Sarcoma 180 in doses of 25 mg/kg resulted in no significant effects (see Fig. 2).

By chemical modifications of the genuine glucans it was hoped to obtain more effective derivatives.

For investigating the influence of ß-1.6 ramnifications on the antitumor activity of glucans, 6-0-glucosyl derivatives of cellulose, amylose and barley glucan were synthesized according to Husemann et al. (9). The degree of branching was about 45%. None of these derivatives exhibited with doses ranging from 1 to 5 mg/kg significant inhibition rates.

Synthesis of further derivatives such as the glucan ethers and -esters was carried out by reactions of the glucans with dialkyl sulfates, chloracetic acid, carbonic acid anhydrides or -chlorides and chlorosulfonic acid.

All derivatives were tested at Sarcoma 180 in doses of 25mg/kg.

All semisynthetic derivatives of barley glucan exhibited a higher activity (Fig. 1) as the unsubstituted polysaccharide. In contrast, the genuine oat glucan showed highest inhibition rates, which were not enhanced by the introduction of side chains.

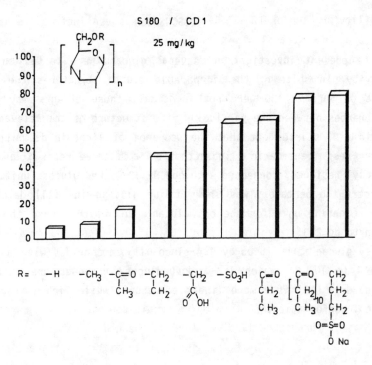

Figure 1: Inhibition rates of various Barley glucan
derivatives.

DISCUSSION

In the present investigations several glucans were tested on antitumor activity. In spite of the comparable glucan structures the inhibition rates on Sarcoma 180 were not identical. These effects may be due to differences between the particular fine structure of the relevant glucans. Perlin (10) pointed out that the sequence of lichenin is built up by a tetra- resp. pentameric oligosaccharide with three resp. four 1.4-linked glucosyl residues, connected with one ß-1.3-linked glucose unit. The ratio of tetra- to pentamers was shown to be 4:1. Nevins (11) determined the same repeating unit for the oat glucan, but in this case the ratio was shown to be 2,3:1.

Barley glucan is built up by 1.4-gluco-oligosaccharides with a DP ranging from 4 to 10, to which 1.3-linked glucose residues are attached. The consequence (12) is a more linear structure, while lichenin has a strong helical conformation.

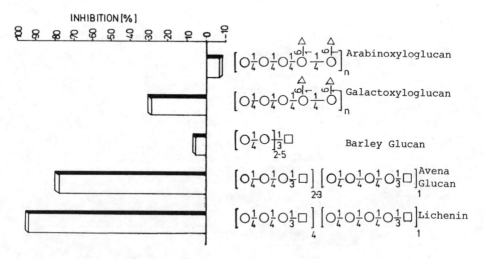

Figure 2: Dependence of antitumor activity in correlation with the fine structure of glucans

Figure 2 shows the dependence of antitumor activity in relation to the glucan conformation. As a result one can see, the greater the helical proportion of the glucans, the higher the tumor inhibition rate.

As shown with isolichenin, α-configurated glucans are less effective than ß-linked polysaccharides.

Linear glucans with a ß-1.4-linked backbone have no antitumor effect at all.

Looking at the results obtained with the different glucan derivatives, we can see that relations between structure and effects must be investigated for every polysaccharide. General statements can not be given because of the different fine structures of these polymers. As shown with barley glucan derivatives, substitution with long side chains or very polar groups enhanced the activity. Combination of large and acidic substituents (e.g. propoxy-sulfo-groups) are most active.

Higher proportions of ß-1.3-linked glucose residues in the polymers led to higher inhibition rates.

Because of the cheap and easy preparation of the glucan derivatives, such substances could gain an important value in cancer treatment.

ACKNOWLEDGEMENT

The authors obtained financial support from the 'Deutsche Krebshilfe' and from the 'Fonds der Chemischen Industrie'.

REFERENCES

1. Chihara, G., Immunopharmacology of Lentinan and the Glucans. EOS Riv. Immunol. Immunfarmacol., 1984, **4** (I) 2a, 85-95.

2. Matsuo, T., Arika, T., Mitani, M. and Komatsu, N., Pharmacological and Toxicological Studies of a New Antitumor Polysaccharide, Schizophyllan. Drug Res., 1982, **32** (I):6, 647-656.

3. Kraus, J., Schneider, M. and Franz, G., Antitumorpolysaccharide. Deutsch. Apothekerzeitung, 1986, **38**, 2045-49.

4. Diller, I.C.,Mankowski, Z.T. and Fischer, M.E., The Effect of Yeast Polysaccharides on Mouse Tumors. Cancer Res.,1963, **23**, 201-208.

5. Fujiware, T., Takeda, T., Ogihara, Y., Shimizu, M., Nomura, T. and Tomity, Y., Studies on the Structure of Polysaccharides from the Bark of Melia azadirachta. Chem. Pharm. Bull., 1982, **30**, 4025-4030.

6. Whistler, T.L., Bushway, A.A. and Sing, P.P., Noncytotoxic Antitumor Polysaccharides. Adv. Carbohydr. Chem. Biochem., ed. R.S. Tipson, D. Horta, 1976, **32**, 235-280.

7. Akiyama, Y., and Kato, K., An Arabinoxyloglucan from Extracellular Polysaccharides of Suspension-Cultured Tobacco Cells. Phytochem., 1982, **21:8**, 2112-14.

8. Hsu, D.S. and Reeves, R.E., The Structure of Nasturtium Amyloid. Carbohydr. Res., 1967, **5**, 202-209.

9. Husemann, E. und Reinhardt, M., Über die Darstellung definiert verzweigter Polysaccharide (I). Makromol. Chem., 1962, **57**, 109-128.

10. Perlin , A. and Suzugi, S., The Structure of Lichenin: Selective Enzymolysis Studies. Can. J. Chem., 1962, **40**, 50-56.

11. Yamamoto, R. and D.J. Nevins, Structural Studies on the ß-D-Glucan of the Avena Coleoptile Cell Wall. Carbohydr. Res., 1978, **67**, 275-280.

12. Buliga, G.S. , Brant, D..A. and Fincher, G.B., The Sequence Statistics and Solution Conformation of a Barley (1.3, 1.4)-ß-D-Glu- can. Carbohydr. Res., 1986, **157**, 139-156.

13. Wada, S. and Ray, P., Matrix Polysaccharides of Oat Coleoptile Cell Wall. Phytochem., 1978, **17**, 923-931.

14. Peal, S. Whelan, W.J., Turvey, J.R. and Morgan, K., The Structure of Isolichenin. J., 1961, 623-629.

15. Chihara, G., Hamuro, J., Maeda, Y., Arai, Y. and Fukuoka, F., Antitumor Polysaccharide derived Chemically from Natural Glucan (Pachyman). Nature, 1970, **255**, 943-944.

16. Helberger, J.H., Manecke, G. and Heyden, R., Zur Kenntnis organischer Sulfonsäuren. Analen 1949, **565**, 22-35.

17. Hämmerling, U. and Westphal, O., Synthesis and Use of O-Stearoyl Polysaccharides in Passive Hemagglutination and Hemolysis. Europ. J. Biochem. 1967, **1**, 46-52.

18. Cavalli-Sforza, L., Biometrie. G. Fischer, Stuttgart 1969.

DISCUSSION

Q: R. M. Brown, Jr. (Univ. Texas)
(1) Has the anti-tumor agent been tested with humans ? If so, what are the prospects ?
(2) When the tumors· have repressed, how long can the animals maintain a tumor-free state ?

A: (1) Clinical tests (Phase I and II) are being in progress since no side effects have been observed whatsoever. The most effective polysaccharides and their derivatives will only be used against micrometastasis formation and not against solid human tumors. Application will as well be done in combination with chemotherapy and radiation therapy.
(2) Antitumor treatment can be repeated but in much cases the tumors did not reappear.

Q: Y. Tezuka (Tech. Univ. Nagaoka)
Did you observe any pharmacological effect by changing counter cation on sulfonate-saccharide (lichenin) from sodium to others, like lithium or potassium ?

A: This has not been done, but is an excellent suggestion for a possible alternation of the physiological effect of the substituted lichenin derivative.

Q: K. Matsuzaki (Emer. Prof. Univ. Tokyo)
Did you change the substitution of γ-propoxy sulfate and how about the antitumor effect ?

A: An optimal substitution (γ-propoxy sulfonate derivative) was obtained with an 8% sulfur content and a regular DS between 6 and 8.

Q: J. Sunamoto (Kyoto Univ.)
Is the triple helices structure of polysaccharide a basic requirement in their physiological activity ? Or is it only requirement for increasing by macrophage uptake ?

A: It is not clear if the triple helix is an obligatory for requisite of any immune modulating activity: schizophyllan and lentinan are forming triple helices ; lichenin and its derivatives are active but are not forming any triple helix, macrophage activity is increased in each case.

SELECTIVE SYNTHESIS OF BIOLOGICALLY ACTIVE POLYSACCHARIDES

TOSHIYUKI URYU, KENICHI HATANAKA, TAKASHI YOSHIDA, KEI MATSUZAKI,* YUTARO
KANEKO,[+] TORU MIMURA,[+] HIDEKI NAKASHIMA,[+] OSAMU YOSHIDA,[+] and
NAOKI YAMAMOTO[+]
Institute of Industrial Science, University of Tokyo, Roppongi, Manato-ku,
Tokyo 106, *Sinshu University, Ueda, Nagano 386, [+]Ajinomoto Co., Inc.,
Chuo-ku, Tokyo 108, and [+]Department of Virology and Parasitology,
Yamaguchi University School of Medicine, Ube, Yamaguchi 755, Japan

ABSTRACT

It is reported that in vitro anti-AIDS virus and anticoagulant activities
of sulfated, linear polysaccharides, and sulfated branched ones were assayed and a highly potent inhibitor of AIDS-virus infection was successfully
synthesized by sulfation of a natural branched polysaccharide, lentinan.
Lentinan sulfate completely inhibited the infection of AIDS virus to T
lymphocytes in the drug concentration of as low as 3.3 µg/ml.

INTRODUCTION

Heparin is a polysaccharide containing such anionic groups as sulfamide,
sulfate, and carboxyl groups, and has a strong anticoagulant activity. The
function originates from the biologically active pentasaccharide portion
which binds to an anticoagulant factor, antithrombin III, in the blood [1].
Applying the ring-opening polymerization of anhydro sugars [2], we synthesized polysaccharides containing amino groups by polymerizing an anhydro
sugar having an azido group [3]. The anticoagulant activity of the sulfated
polysaccharides containing the sulfamide and sulfate groups were measured
in vitro and in vivo [4]. Moreover, we have revealed that a high anticoagulant activity is shown by sulfated synthetic polysaccharides, especially
furanan-type polysaccharides.

Recently, it was suggested that dextran sulfate has a unique biological
activity, i.e., an inhibitory effect on AIDS virus (= human immunodeficiency virus, HIV) infection by inhibiting the activity of its reverse transcriptase [5]. When dextran sulfate is used with azidothymidine (= zidovudine),

it exhibits the inhibitory effect on the virus infection in vitro [6].

We have found that the sulfated synthetic polysaccharides strongly inhibit the infection of AIDS virus in vitro [7]. In order to synthesize a sulfated polysaccharide with a higher biological activity, a natural branched 1,3-glucan, lentinan, was sulfated [8].

EXPERIMENT

Compounds. Sulfamide group-containing polysaccharides were prepared by sulfating 3-amino-3-deoxy-(1→6)-α-D-glucopyranan and its copolymer with chlorosulfonic acid [3,4]. Synthetic polysaccharides were obtained by polymerization of 1,4-anhydro sugars [9,10]. Sulfation of the free poly-saccharides was carried out with piperidine N-sulfonic acid in dimethyl-sulfoxide [5]. Dextran used for sulfation was kindly provided from Northern Regional Research Laboratory, United States Department of Agriculture. Lentinan was supplied by Ajinomoto Co., Inc. Dextran sulfate (NC-1032) ($\overline{M}_W$ 7000) was provided by Meito Sangyo, Ltd.

Measurements. Anticoagulant activity was measured in vitro using bovine serum according to a modification of United States Pharmacopoeia [4].

Anti-HIV assay. Anti-HIV activity of sulfated polysaccharides was assayed by measuring the decrease in the number of viable cells and indirect immuno-fluorescence (IF) method using MT-4 cells [11].

Inhibition of syncytium formation. To examine multinucleated giant cell formation which is caused by cell-to-cell infection, a human T-cell line, MOLT-4 and its HIV-production cell, MOLT-4/HIV$_{HTLV-IIIB}$ were employed [12]. The two cells were mixed in a ratio of 1:1 adjusting to a final cell density of 5×10^5 cells/ml.

Reverse transcriptase (RT) inhibition assay. The RT inhibitory activity of sulfated polysaccharides were measured at 37°C for 30 min by mixing 1U of purified avian myeloblastosis virus (AMV) reverse transcriptase with various concentrations of the compounds [13]. HIV transcriptase was also used [7].

RESULTS

1. Anticoagulant activity of sulfated polysaccharides.

The anticoagulant activity of sulfated synthetic polysaccharides is shown in TABLE 1. Sulfated dextran-type polysaccharides with different

sulfamide contents show   anticoagulant activities of 35 and 41 units/mg,
indepedent of the sulfamide content.

The synthesis of higher anticoagulant compounds was investigated. 1,4-
Anhydro-α-D-xylopyranose (= 1,5-anhydro-β-D-xylofuranose) derivative is
polymerized to give (1→5)-α-D-xylofuranan and an isomeric poly-D-xylose
with a mixed structure consisting of 1,5-α-D-xylofuranose and 1,5-β-D-xylo-
furanose units [10]. The stereoregular (1→5)-α-D-xylofuranan was sulfated
with piperidine N-sulfonic acid to produce xylofuranan sulfate. Next, 1,4-
anhydro-α-D-ribopyranose was polymerized into three isomeric polysaccharides, i.e., (1→5)-α-D-ribofuranan, (1→4)-β-D-ribopyranan, and a nonstereoregular poly-D-ribose [9]. The poly-D-riboses were sulfated by the same
method as that of the poly-xylose.

As shown in TABLE 1, stereoregular (1→5)-α-D-xylofuranan (No. 7) and
-ribofuranan sulfates (No. 10) exhibited very high anticoagulant activities, 69 and 56 units/mg, respectively. (1→4)-β-D-Ribopyranan sulfate (No.
11) had a little lower activity (41 units/mg) than that of the isomeric
ribofuranan sulfate. Nonstereoregular sulfated polysaccharides showed low

TABLE 1. Anticoagulant Activity in vitro and LD_{50} of Sulfated
Polysaccharides

No.	Polysaccharide type	Sulfamide content[a]	Sulfur content %	$\bar{M}n \times 10^{-4}$	Anticoagulant activity(A.A) units/mg	LD_{50} mg/kg
1	(1→6)-α-D-glucan	1.0	16.1	1.45	35	1324
2	(1→6)-α-D-glucan	0.5	14.7	2.01	41	710
3	(1→6)-α-D-glucan	0	——	2.83	34	518
4	dextran sulfate NC-1032	0	18.4	0.70	20.6	2500
5	(1→6)-α-D-glucan branched (B-742)	0	11.2	2.1 ($\bar{M}_W$ 3.4)	74	——
6	(1→6)-α-D-glucan mannose(90%)-branched	0	——	——	100	——
7	(1→5)-α-D-xylofuranan	0	16.1	1.24	69	——
8	(1→5)-α-D-xylofuranan	0	15.7	1.0	28	——
9	(1→5)-α-D-xylofuranan mannose(10%)-branched	0	16.3	0.7	50	——
10	(1→5)-α-D-ribofuranan	0	15.5	0.89	56	——
11	(1→4)-β-D-ribopyranan	0	15.1	3.48	41	——

a: Number of sulfamide groups per sugar unit.

activities (data not shown). Consequently, it was revealed that furanan-type polysaccharides have higher activities.

The 6-O-sulfate group in the active pentasaccharide portion of heparin contributes to a large extent to its high activity [14]. Thus, branching of the linear polypentose, xylofuranan, was performed to introduce a sulfate group at the C-6 position. This was achieved by copolymerization of 1,4-anhydro-2,3-di-O-tert-butyldimethylsilyl-α-D-xylopyranose with a 2,3-di-O-benzylated homologue, followed by selective removal of silyl group and by branching with mannose orthoester [15]. Comparing the anticoagulant activity (28 units/mg) of No. 8 xylofuranan sulfate ($\overline{M}_n$ 10 x 10^3) with that (50 units/mg) of No. 9 mannose-branched xylofuranan sulfate ($\overline{M}_n$ 7 x 10^3), the branching and the sulfation at the C-6 position of the mannose branch caused an increase in the activity. High activities were observed in a natural branched dextran (No. 5)(AA 74 units/mg) and a synthetic mannose-branched dextran (No. 6)(AA 100 units/mg).

Concerning the toxicity, LD$_{50}$, measured by use of mice, polysaccharides containing higher sulfamide contents showed lower toxicities, revealing that the sulfamide group has a function to decrease toxicity in vivo.

2. Anti-HIV (Anti-AIDS virus) activity of linear sulfated polysaccharides.

It was reported that sea algae and other natural sulfate-containing polysaccharides efficiently inhibit HIV infection and virus-induced cytopathic effect as well as reverse transcriptase activity [16]. The anti-HIV activity was examined on dextran sulfate ($\overline{M}_w$ 7 x 10^3), synthetic xylofuranan sulfate, and synthetic ribofuranan sulfate. Recently, a natural linear (1→3)-β-D-glucan, curdlan, was successfully sulfated to give curdlan sulfate.

Anti-HIV activity of the sulfated compounds was assayed on the prevention of HIV-induced cytopathic effects (CPE), i.e., on the number of viable cells, and the expression of virus-specific antigens on the surface of cells. After the virus adsorption treatment, the cells were washed and adjusted to a concentration of 3 x 10^5 cells/ml.

The effects of sulfated polysaccharides on growth of MT-4 cells and on HIV-induced cytopathic effects were evaluated on the 3rd and 6th days. The results are summarized in Figure 1. Concentrations of more than 10 µg/ml of xylofuranan and ribofuranan sulfates completely protected MT-4 cells from HIV-induced CPE, letting the cells grow to a concentration of about 25 x 10^5 cells/ml. Dextran sulfate ($\overline{M}_w$ 7 x 10^3) had a little lower anti-HIV activity than the above two compounds, exhibiting complete inhibitory effects at concentrations of more than 100 µg/ml. All these sulfated polysaccharides showed no cell-growth inhibition at the concentration of 1000 µg/ml.

The proportion of HIV-infected cells was estimated by indirect immunofluorescence method. The results are shown in Figure 2. Of the MT-4 cells which were incubated in the presence of xylofuranan and ribofuranan sulfates in the concentration of 10 µg/ml, about 10% and 25% cells had HIV-specific antigens, respectively, while almost no antigen appeared at the concentration of 100 µg/ml.

Curdlan sulfate had potent inhibitory effects on the HIV-infection at the concentration of 3.3 µg/ml. At this concentration, the incubated cells showed almost no expression of HIV-specific antigens.

For dextran sulfate, xylofuranan sulfate, and ribofuranan sulfate, the inhibitory effects on reverse transcriptase activities of AMV and HIV were in the range of 90 - 96% in the concentrations of more than 100 µg/ml, except that the inhibitory effects of dextran sulfate ($\overline{M}_W$ 7 x 10^3) on the

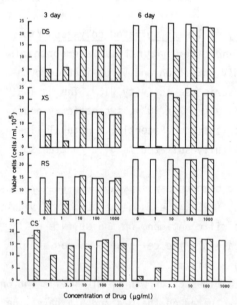

Fig. 1. Effects of sulfated linear polysaccharides on inhibition of HIV infection and cell growth. DS:Dextran sulfate (M_W 7000); XS:Synthetic xylofuranan sulfate; RS:Synthetic ribofuranan sulfate; CS:Curdlan sulfate.

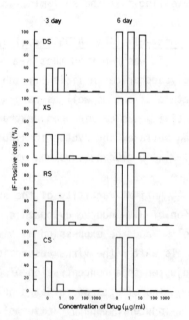

Fig. 2. Inhibitory effects of sulfated polysaccharides on the expression of HIV-specific antigen as determined by indirect immunofluorescence method.

RT of HIV was 47%.

In the coculture of MOLT-4 cells and MOLT-4/HIV$_{HTLV-IIIB}$, multinucleated giant cells had appeared by 1 day after the culture started. When 100 μg/ml of dextran sulfate, xylofuranan sulfate, or ribofuranan sulfate were added to the coculture medium, the formation of giant cells was completely inhibited in the 6 days observation.

3. Anti-HIV activity of sulfated branched polysaccharides.

It was intended to prepare sulfated polysaccharides having high anti-HIV activity but low anticoagulant activity, because the anticoagulant activity is a side reaction for anti-HIV active compounds. Sulfation of branched polysaccharides might have a possibility to produce polysaccharides with the above properties. Natural dextran with a branched structure (NRRL B-742)(1→6:70%, 1→4:13%, 1→3:18%) was sulfated to give dextran sulfate ($\overline{M}_W$ 34 x 10^3), which had a high anticoagulant activity of 74 units/mg (TABLE 1). This dextran sulfate was tested on the anti-HIV activity. As shown in Figure 3, the dextran sulfate showed complete inhibitory effects on the HIV infection in concentrations of more than 10 μg/ml, being similar to linear xylofuranan and ribofuranan sulfates. However, 20 - 30% cell growth inhibition was observed in the concentration of 1000 μg/ml of the dextran sulfate.

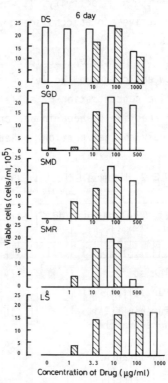

Glucose- and mannose-branched dextrans, and a mannose-branched ribofuranan were synthesized. All three synthetic branched polysaccharides were sulfated to give sulfated polysaccharides with strong inhibitory effects on HIV infection in the concentration of more than 10 μg/ml. In this concentration, the cells contained almost no IF-

Fig. 3. Effects of sulfated branched polysaccharides on inhibition of HIV infection and cell growth. DS:Dextran sulfate (M_W 34000); SGD:Synthetic glucosyl-dextran sulfate; SMD: Synthetic mannosyl-dextran sulfate; SMR:Synthetic mannosyl-ribofuranan sulfate; LS:Lentinan sulfate.

positive ones.

Lentinan which is an antitumor polysaccharide having a structure of branched (1→3)-β-D-glucopyranan [17] was sulfated. Lentinan sulfate showed the strongest anti-HIV activity, that is, complete inhibitory effects in the concentration of 3.3 μg/ml. In this concentration, the cultured cells contained almost no IF-positive ones. In addition, the lentinan sulfate had relatively low anticoagulant activity of 21 units/mg.

DISCUSSION

In the early stage of this investigation, it was aimed to synthesize a highly anticoagulant polysaccharide. Branched dextran sulfates were found to have the high activity. It was based on the finding that the 6-O-sulfate in the active pentasaccharide portion of heparin acts to enhance the activity [15].

In the next step, sulfated polysaccharides with high anti-HIV activity but low anticoagulant activity have successfully been prepared. The minimum effective concentration of sulfated polysaccharides is shown in TABLE 2. In the reaction of heparin and antithrombin III, it was proposed that the negatively charged portion in heparin binds with the positively charged

TABLE 2. Minimum Effective Concentration of Sulfated Polysaccharides on Complete Inhibition of HIV Infection

Sulfated Polysaccharide	S content %	$\bar{M}_n \times 10^{-3}$	Minimum effective μg/ml
Lentinan sulfate	16.2	18	3.3
Curdlan sulfate	14.5	125	3.3
Dextran sulfate		34	10
Ribofuranan sulfate (Mannose branched)	15.3		10
Dextran sulfate (glucose branched)	14.4	47	10
Dextran sulfate (Mannose branched)	15.5	53	10
Ribofuranan sulfate (Linear)	15.7	10	10
Xylofuranan sulfate (Linear)	14.2	7	10
Dextran sulfate	18.4	7	100

lysine residues [18]. Thus, it is reasonable to consider that the sulfated polysaccharides' having many negative charges react with accumulated lysine residues in surface glycoproteins of HIV, preventing the virus from fusing on T cells. Since the anti-HIV activity increased with increasing molecular weight of sulfated polysaccharides, it might be considered that they work outside of the virus and the T cell, but not inside of the T cell. Of course, the sulfated polysaccharide can inhibit the reverse transcriptase activity, if it enters inside of the virus and the T cell.

REFERENCES

1. Björk, I. and Lindahl, U., Mol. Cell. Biochem., 1982, 48, 161.
2. Ruckel, E. R. and Schuerch, C., J. Am. Chem. Soc., 1966, 88, 2605.
3. Uryu, T., Hatanaka, K., Matsuzaki, K. and Kuzuhara, H., Macromolecules, 1983, 16, 853.
4. Hatanaka, K., Yoshida, T, Miyahara, S., Sato, T., Ono, F., Uryu, T. and Kuzuhara, H., J. Med. Chem., 1987, 30, 810.
5. Diringer, H. and Mehring, K., Japan Kokai, S62-215529, 1987.
6. Ueno, R. and Kuno, S., Lancet, 1987, June 13, 1379.
7. Nakashima, H., Yoshida, O., Tochikura, T. S., Yoshida, T., Mimura, T., Kido, Y., Motoki, Y., Kaneko, Y., Uryu, T. and Yamamoto, N., Jpn. J. Cancer Res. (Gann), 1987, 78, 1164.
8. Yoshida, O., Nakashima, H., Yoshida, T., Kaneko, Y., Yamamoto, I., Matsuzaki, K., Uryu, T. and Yamamoto, N., Biochem. Pharm., 1988, 37, 2887.
9. Uryu, T., Yamanouchi, J., Kato, T., Higuchi, S. and Matsuzaki, K., J. Am. Chem. Soc., 1983, 105, 6865.
10. Uryu, T., Yamanouchi, J., Hayashi, S., Tamaki, H. and Matsuzaki, K., Macromolecules, 1983, 16, 320.
11. Hamamoto, Y., Nakashima, H., Matsui, T., Matsuda, A., Ueda, T. and Yamamoto, N., Antimicrob. Agents Chemother., 1987, 31, 907.
12. Nakashima, H., Tochikura, T., Kobayashi, N., Matsuda, A., Ueda, T. and Yamamoto, N., Virology, 1987, 159, 169.
13. Harada, S., Koyanagi, Y. and Yamamoto, N., Virology, 1985, 146, 272.
14. Atha, D. H., Lormeau, J. C., Petitou, M., Rosenberg, R. D. and Chouy, J., Biochemistry, 1985, 24, 6723.
15. Yoshida, T., Arai, T., Mukai, Y. and Uryu, T., Carbohydr. Res., 1988, 177, 69.
16. Nakashima, H., Kido, Y., Kobayashi, N., Motoki, Y., Neushul, M. and Yamamoto, N., J. Cancer Res. Clin. Oncol., 1987, 113, 413.
17. Chihara, G., Maeda, Y. Y. and Hamuro, J., Int. J. Tiss. Reac., 1982, IV, 207.
18. Villanueva, G. B., J. Biol. Chem., 1984, 259, 2531.

DISCUSSION

Comment: K. Iiyama (Univ. Tokyo)

Lignosulfonate (Ca^{++}, Na^+) also show the immunopotentiating activity by the cheques with macrophage activity. The immunopotentiating activity may be introduced by the polyanion with three dimensional network polymer.

Further, the activity may depend on the molecular weight and content of $-SO_3-$ group of lignosulfonate.

The above results were submitted to Agric. Biol. Chem. as a preliminary paper.

A: I showed the influence of molecular weight and sulfate content on the anti-AIDS virus activity in the slide. Both are very important to acquire high activity. Concerning lignosulfonate, I think that the component might have high toxicity.

HYALURONAN, ITS CROSSLINKED DERIVATIVE -- HYLAN -- AND THEIR MEDICAL APPLICATIONS

Endre A. Balazs, M.D. and Edward A. Leshchiner, Ph.D.
Biomatrix, Inc.,
65 Railroad Avenue, Ridgefield, NJ 07657 USA

ABSTRACT

Hyaluronan (hyaluronic acid, HA) in its purest form has been introduced in human and veterinary medicine as a viscosurgical tool in ophthalmology and as a treatment for traumatic arthritis in horses and humans. This natural polysaccharide has unusual rheological properties and excellent biocompatibility. The native HA molecule was crosslinked using two distinctly different methods, and the resulting molecular structures are called hylans (generic name). Hylans can be prepared in water-soluble form, and the molecules, having 8-23 million molecular weight, form extremely elastoviscous solutions. Hylans also exist in water-insoluble forms, when the crosslinks between the polysaccharide chains produce an infinite network (gels, membranes, tubes, etc.). Hylan gels have been shown to be highly biocompatible: non-antigenic, non-inflammatory, non-tissue reactive. Hylans are intended to be used in medicine as viscosurgical tools and implants. Hylans are also intended to be used for viscosupplementation of the joints and other tissues. They are usable in matrix engineering as matrices to provide space-filling implants engineered to accomplish specific therapeutic goals. The implanted matrices direct and control wound healing and regenerative processes in the soft connective tissues. Hylan gels and solids are also usable as delivery systems for therapeutic agents.

INTRODUCTION

During the 1980's hyaluronan and its derivatives (hylans) became important medical devices and implants. They are used in a variety of medical applications such as ophthalmology, orthopedic surgery, rheumatology, otology, plastic surgery and veterinary medicine. In this paper their chemical structure, rheological properties and medical use will be discussed. There are two properties which make hyaluronan and hylans, more than other natural polysaccharides, the most suitable molecules for medical use: first, the unusual rheological properties of the native hyaluronan that were further enhanced by broadening the physical form and properties of its crosslinked derivatives, the hylans; and second, the remarkable biological compatibility of these polysaccharides.

CHEMICAL STRUCTURE

Hyaluronan (hyaluronic acid, HA or hyaluronate) is a glycosaminoglycan with a repeating disaccharide unit of N-acetylglucosamine and glucuronic acid. These two sugars are linked with a ß1 → 4 glucoside bond, and the repeating disaccharide units are linked with ß1 → 3 glucoside bonds. The unbranched chain forms a random coil with some stiffness, having a weight average molecule weight of 5-6 million [1].

Hylan is the generic name for crosslinked hyaluronan chains when the crosslinking does not affect the two specific groups of the molecule, namely the carboxylic and N-acetyl groups. Hylans have the same polysaccharide chain and polyanionic characteristics as hyaluronan, but either their molecular weight is higher or they form infinite molecular networks (gels). Hylans are obtained by two crosslinking processes. In the first, formaldehyde is used at neutral pH to produce a permanent bond between the C-OH group of the polysaccharide and the amino or imino groups of a protein [2]. This protein forms a bridge between two polysaccharide chains. To establish this crosslink, the polysaccharide chains must be the correct distance from each other, and the protein forming the crosslink must have a proper relationship to the polysaccharide. Only specific proteins of relatively small molecular size and with specific affinity to the hyaluronan chain will satisfy the definition of bridge proteins. Under the right conditions, the crosslinking process will yield not an infinite molecular network, but only a permanent association of two to eight hyaluronan molecules. Thus the weight average molecular weight of this hylan molecular population will be 8-24 million. The protein content of such a purified hylan will be 0.4-0.8% of the polysaccharide content. Since the bridge protein is buried between two or more hyaluronan chains, it does not act as an antibody. This crosslinking process produces a less dense structure than that of the native hyaluronan, resulting in a partial specific volume for hylan of 0.63 cc/g compared to hyaluronan's 0.57 cc/g (both in 0.15N NaCl). Since only a limited number of the polysaccharide chains of hyaluronan are permanently associated through a methylene--bridge-protein--methylene bond, the hydrated molecules form an elastoviscous solution which is called hylan fluid [3].

The other crosslinking process utilizes vinyl sulfone which reacts with the hydroxyl groups of the polysaccharide chain to form an infinite network through sulfonyl-bis-ethyl crosslinks [4,5]. The crosslinked gel has a very high hydration level (≤ 99.7% water) but this can be varied to form more dense gels by altering the conditions of the crosslinking process. The gel is then broken up into small pieces which are deformable and easily aggregate (reversibly) into larger gel pieces. The slurry of the gel is very viscoelastic, especially when it is made up of highly hydrated small gel pieces. The same sulfonyl-bis-ethyl crosslinks between hyaluronan or hylan chains can form hydrated structures with a 2-10% polysaccharide content. These structures can be formed into sheaths, membranes, tubes, sleeves, and particles of various sizes and shapes.

RHEOLOGY

Hyaluronan. The rheological properties of solutions of native hyaluronan are characterized by extremely shear-dependent viscosity (Figure 1) and frequency-dependent elasticity (Figure 2) [6,7,8]. Hyaluronan solutions are also very pseudoplastic. These rheological properties depend on the concentration and molecular weight of the polysaccharide material.

Hylans. The rheological properties of hylan fluid are different from those of hyaluronans of the same concentration. At the lowest frequencies and shear rates the elasticity and the shear and dynamic viscosities are significantly higher in hylan than in hyaluronan solutions (Figures 1 and 2). This means that the hylan fluids exhibit predominantly elastic properties at lower frequency than hyaluronan solutions. There are two reasons for this: the molecular weight of hylan is greater than that of hyaluronan, and the interaction between the individual molecules and between the solute and the solvent is different for the two polysaccharides.

The rheological properties of slurries made of hylan gel are completely different from hylan fluid. Hylan gel slurries are made up of small gel pieces without a solvent interphase between them. The gel pieces are very deformable and move relative to each other under laminar flow conditions. We therefore characterize the rheological properties of this slurry using the same parameters as the hylan fluid (Figures 1 and 2). The elasticity of the gel slurry at all frequencies and the viscosity at low shear rates are significantly greater than those of hylan fluid.

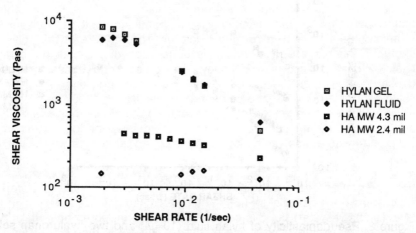

Figure 1. Shear rate dependent viscosity of hylan gel slurry (polymer concentration 4 g/l), hylan fluid (10 g/l) and two hyaluronan solutions (10 g/l) of different molecular weights.

236

Hylan and hyaluronans are pseudoplastic polymers with no real yield point, as shown in Figure 3. On the other hand, hylan gel has plastic properties and exhibits a yield point which depends on the degree of crosslinking and the size of the gel particles of the slurry. But the principal difference between hylan gel, hylan fluid and hyaluronan is the difference in solubility. The gel particles of hylan gel can be separated by and dispersed in hydrophilic fluids, but they remain insoluble in water [5].

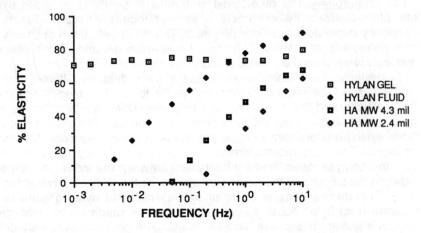

Figure 2. Frequency-dependent elasticity of hylan gel slurry (0.4 g/l), hylan fluid (10 g/l) and two hyaluronan solutions (10 g/l) of different molecular weight polymers.

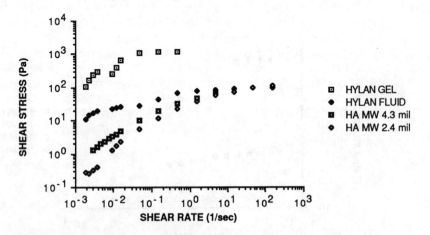

Figure 3. Pseudoplasticity of hylan fluid (10 g/l) and two hyaluronan solutions of different molecular weight polymers, and plasticity of hylan gel slurry (4 g/l).

BIOCOMPATIBILITY AND RESIDENCE TIME

Hyaluronan is a component of all connective tissues of man and most of the other vertebrates. No species variations have been found in the chemical and physical structure of this glycosaminoglycan. It was found that a chemotactic agent can be associated with the large molecules. Only the hyaluronan fraction that is free of the chemotactic agent can be used in veterinary and human medicine (non-inflammatory fraction of Na-hyaluronan; NIF-NaHA) [9]. NIF-NaHA is not antigenic, and when applied into the connective tissue spaces of the human or animal body, it does not elicit inflammatory or foreign-body reactions [10,11]. The residence time of the injected NIF-NaHA depends on the amount, the rheological properties and the tissue compartment to which it is applied (3 days in the horse joint and 60 days in the monkey vitreus) [11]. In all tissues the undegraded molecule diffuses or flows to the lymph vessels and from there to the blood. The endothelial cells of the liver are primarily responsible for its complete intracellular degradation [12,13].

Hylan Fluid has the same biocompatibility as NIF-NaHA [14]. Due to the significantly higher viscosity of hylan at low shear (≤ 0.01 sec^{-1}) its residence time in connective tissue compartments is considerably longer than that of the highest molecular weight hyaluronan.

Hylan Gels do not elicit antibody formation and they are not toxic to cells and tissues of the body [14]. They do not cause local inflammatory reactions, foreign-body reactions or encapsulation. Therefore they are ideal implants for tissue separation, supplementation and augmentation. The residence time of hylan gel slurries depends on the hydration level of the gel, the size of the gel particles, the elastoviscosity of the fluid separating the gel particles and, most importantly, on the nature of the connective tissue compartment in which it resides. In a tissue compartment which is completely immobile and whose fluid content does not flow (like the vitreus of the eye), the hylan gel remains unaltered and undegraded for an indeterminate amount of time [15,16]. On the otherhand, the half-life time of the same gel slurry in a tissue compartment where constant movement creates considerable pressure and fluid flow is only 20-30 days. Most of the joints of the body are examples of such tissue compartments. The small particles of the gel slurry under the influence of the movement of the joint and the flow of joint fluid migrate to the soft connective tissues of the joint capsule. During this migration the small gel particles further break down mechanically and enter the lymph vessels, passing through the lymph nodes to the blood. From there it is removed by the liver endothelium and degraded by the same mechanism as the hyaluronan molecules.

MEDICAL APPLICATION

Hyaluronan (NIF-NaHA) and hylans are used in medicine as medical tools and implants because of their unique physical properties. The mechanism of their action is physical and not chemical. Their mode of action and utility in medicine can be described in four areas:

Viscosurgery. Elastoviscous surgical tools or implants are used to help carry out the objectives of surgery [17,18]. During the surgical procedure the viscosurgical tools separate tissues and provide space for surgical manipulation. They serve as tissue protectors and lubricators for mechanical tools, thereby preventing damage to sensitive cell layers. They themselves serve as soft surgical instruments to move tissues, clear tissue debris, form a barrier to blood flow and separate tissue adhesions. Elastoviscous viscosurgical implants left at the site of surgery exclude blood cells and fibrinogen, thereby decreasing postoperative bleeding, exudation, scar tissue formation and adhesions. They provide barriers between tissue surfaces, insuring separation of tissues necessary for normal function.

Hyaluronan (NIF-NaHA) was first used as a viscosurgical tool and implant in ophthalmic surgery [19,20]. Recently, clinical studies were initiated to use hylan fluid and hylan gel as viscosurgical tools or implants in those surgical cases when hyaluronan solutions did not provide appropriate rheological properties for viscosurgery. Examples of such cases are glaucoma surgery, vitreo-retinal surgery, trauma surgery and surgery on extraocular muscles.

Viscosurgical principles are being investigated in arthroscopic surgery of large and small joints [21]. Hylan fluid is used to prevent mechanical damage (scuffing) by conventional instruments on the surface of articular cartilage, and to control bleeding, movement of tissues and tissue debris during surgery. Postoperative viscosurgical implants help the normal healing of the tissues of the joint. Hylan gel slurries are used to prevent postoperative adhesions in tendon and joint surgery. While viscosurgical implants do not interfere with the healing of tendons and ligaments, they provide a mechanical separation between surfaces which must remain unattached in order to function normally [22,23]. The use of elastoviscous solution as viscosurgical implants in middle ear surgery is under investigation [24].

Viscosupplementation. In some diseases the normal viscolelasticity of a solid tissue or the elastoviscosity of a tissue fluid is decreased. In such cases hylan fluid or gel slurries can be introduced to restore the normal rheological status of the tissue. Viscosupplementation means that there is restoration of the normal rheological state of the tissue. Under the protective umbrella of the applied viscosupplemental device the normal healing can proceed [25].

Clinical studies indicate that elastoviscous hyaluronan solutions (NIF-NaHA) can restore normal joint fuction in equine traumatic arthritis and in some cases of human chronic idiopathic arthritis [26,27,28,29,30]. Preliminary studies in both fields suggest that hylan fluid and gels with superior rheological properties and longer residence time are more effective in such cases.

Matrix engineering. The development of hylan gels and hylan solids (membranes, tubes, etc.) opened a new field in the medical applications of water-insoluble but high water-containing biocompatible implants [31]. These highly biocompatible polysaccharide structures can serve as scaffolding in directing and controlling tissue regeneration, for filling of tissue spaces, for remodelling of tissue compartments and reshaping of tissue contours. Since these implants are not altered by the immunological defence mechanisms of the body and do not elicit foreign body reactions, they greatly out-perform any other materials (natural or synthetic) used for the above purposes. Matrix engineering is in the preclinical, tissue and animal modelling stage. Its potential usefullness in neurosurgery, plastic surgery and dermatology has been clearly demonstrated.

239

Delivery compartments. Hylan gels and solids can be used as new tissue compartments installed on the surface of the body, on surface wounds or parenterally to deliver therapeutic agents to the neighbouring tissues and to the blood. These delivery compartments made of hylans have the advantage that they do not elicit immunological and foreign body reactions or capsule formation [32].

REFERENCES

1. Balazs, E.A., Physical chemistry of hyaluronic acid. Fed. Proc. 1958, **17,** 1086-1093.

2. Balazs, E.A., Leshchiner, A., Leshchiner, A. and Band, P., Chemically modified hyaluronic acid preparation and method of recovery thereof from animal tissues. United States Patent #4,713,448, 1987.

3. Balazs, E.A., Leshchiner, A., Wedlock, D., Cowman, M., Band, P., Larsen, N., Leshchiner, A. and Hoefling, J., Chemistry, physical chemistry and rheology of hylans. Biomatrix Report #101; Biomatrix, Inc., Ridgefield, New Jersey, USA, 1988.

4. Balazs, E.A. and Leshchiner, A., Hyaluronate modified polymeric articles. United States Patent #4,500,676, 1985.

5. Balazs, E.A. and Leshchiner, A., Cross-linked gels of hyaluronic acid and products containing such gels. United States Patent #4,582,865, 1986.

6. Balazs, E.A., Viscoelastic properties of hyaluronic acid and biological lubrication. Symposium: Prognosis for Arthritis: Rheumatology Research Today and Prospects for Tomorrow, Ann Arbor, Michigan, 1967, Univ. Mich. Med. Ctr. J. (Suppl.). 1968, 255-259.

7. Gibbs, D.A., Merrill, E.W., Smith, K.A. and Balazs, E.A., The rheology of hyaluronic acid. Biopolymers. 1968, **6,** 777-791.

8. Balazs, E.A. and Gibbs, D.A., The rheological properties and biological function of hyaluronic acid. In Chemistry and Molecular Biology of the Intercellular Matrix. ed., E.A. Balazs, Academic Press, London and New York, 1970, pp. 1241-1254.

9. Balazs, E.A., Ultrapure hyaluronic acid and the use thereof. United States Patent #4,141,973, 1979.

10. Denlinger, J.L. and Balazs, E.A., Replacement of the liquid vitreus with sodium hyaluronate in monkeys. I. Short term evaluation. Exp. Eye Res. 1980, **31,** 81-99.

240

11. Denlinger, J.L., Metabolism of sodium hyaluronate in articular and ocular tissues. Ph.D. thesis, Université des Sciences et Techniques de Lille, Lille, France, 1982.

12. Fraser, J.R.E., Laurent, T.C., Pertoft, H. and Baxter, E., Plasma clearance, tissue distribution and metabolism of hyaluronic acid injected intravenously in the rabbit. Biochem. J., 1981, **200,** 415-424.

13. Fraser, J.R.E., Laurent, T.C., Engström-Laurent, A. and Laurent, U.G.B., Elimination of hyaluronic acid from the blood stream in the human. Clin. Exp. Pharm. Phys., 1984, **11,** 17-25.

14. Balazs, E.A., Larsen, N., Morales, B., Denlinger, J.L., Dursema, H. and Kling, M., Biocompatibility of hylans. Biomatrix Report #102; Biomatrix, Inc., Ridgefield, New Jersey, USA, 1988.

15. Vadasz, A., Goldman, A. and Balazs, E.A., The use of hylan gel in ophthalmic surgery. Invest. Ophth. Vis. Sci. (Suppl.), 1988, **29,** 440.

16. Vadasz, A., Balazs, E.A., Goldman, A.I., Ghilardi, M.F., Bodis-Wollner, I. and Glover, A., Hylan gel: rheology, biocompatibility and residence time of a new vitreous gel substitute. Proc. VIII Int. Cong. Eye Res., 1988, **5,** 67.

17. Balazs, E.A., Sodium hyaluronate and viscosurgery. In Healon (sodium hyaluronate) A Guide to Its Use in Ophthalmic Surgery. eds., D. Miller and R. Stegmann, John Wiley and Sons, New York, 1983, pp. 5-28.

18. Balazs, E.A., Viscosurgery. Transplantation/Implantation Today, 1985, **2,** 62-64.

19. Balazs, E.A., Freeman, M.I., Klöti, R., Meyer-Schwickerath, G., Regnault, F. and Sweeney, D.B., Hyaluronic acid and the replacement of vitreous and aqueous humor. In Modern Problems in Ophthalmology, Vol. 10 (Secondary Detachment of the Retina, Lausanne, 1970. ed., E.B. Streiff, S. Karger, Basel, 1972, pp. 3-21.

20. Balazs, E.A., The development of sodium hyaluronate (Healon) as a viscosurgical material in ophthalmic surgery. In Ophthalmic Viscosurgery - A Review of Standards, Techniques and Applications. ed., G. Eisner, Medicöpea, Bern, 1986, pp. 3-19.

21. Weiss, C. and Balazs, E.A., Arthroscopic Viscosurgery. Arthroscopy, 1987, **3,** 138.

22. Weiss, C., Levy, H.J., Denlinger, J.L., Suros, J.M. and Weiss, H.E., The role of Na-hylan in reducing postsurgical tendon adhesions. Bull. Hosp. J. Dis. Ortho. Inst., 1986, **46,** 9-15.

23. Weiss, C., Suros, J.M., Michalow, A., Denlinger, J.L., Moore, M. and Tejeiro, W., The role of Na-hylan in reducing postsurgical tendon adhesions: Part 2. Bull. Hosp. J. Dis. Ortho. Inst.. 1987, **47,** 31-39.

24. Laurent, C., Hyaluronan in the middle ear. Umeå University Medical Dissertations New Series Number 211, 1988.

25. Balazs, E.A. and Denlinger, J.L., Clinical uses of hyaluronan. In The Biology of Hyaluronan. Ciba Foundation Symposium #143, eds., D. Evered and J. Whelan, John Wiley & Sons, Chichester, Sussex, 1989, (under publication).

26. Balazs, E.A. and Denlinger, J.L., The role of hyaluronic acid in arthritis and its therapeutic use. In Osteoarthritis: Current clinical and fundamental problems. Proceedings of a workshop held in Paris, 9-11 April, 1984. 1985a, ed., J.G. Peyron, Geigy, pp. 165-174.

27. Peyron, J.G. and Balazs, E.A., Preliminary clinical assessment of Na-hyaluronate injection into human arthritic joints. Path. Biol.. 1974, **22,** 732-736.

28. Rydell, N.W., Butler, J. and Balazs, E.A., Hyaluronic acid in synovial fluid VI. Effect of intraarticular injection of hyaluronic acid on the clinical symptoms of arthritis in track horses. Acta Vet. Scand., 1970, **11,** 139-155.

29. Weiss, C., Balazs, E.A., St. Onge, R. and Denlinger, J.L., Clinical studies of the intraarticular injection of Healon® (sodium hyaluronate) in the treatment of osteoarthritis of human knees. In Seminars in Arthritis and Rheumatism. Vol. 11, ed., J.H. Talbott, Grune and Stratton, New York, 1981, pp. 143-144.

30. Balazs, E.A. and Denlinger, J.L., Sodium hyaluronate and joint function. J. Eq. Vet. Sci., 1985b, **5,** 217-228.

31. Balazs, E.A., Denlinger, J.L., Leshchiner, E., Band, P., Larsen, N., Leshchinger, A. and Morales, B., Hylan: Hyaluronan Derivatives for Soft Tissue Repair and Augmentation. In Proc. Fifth Ind. Conf. on Biotech.. 1988, (under publication).

32. Balazs, E.A., Leshchiner, A. and Larsen, N., Drug delivery systems based on hyaluronan, derivatives thereof and their salts & methods of producing same. United States Patent #4,713,448, 1987.

DISCUSSION

Q: A. Yamauchi (Res. Inst. Polym. & Text. Japan)
How long does crosslinked H.A. stay in vivo, especially in the vitreous body, undegraded by enzymes ?

A: Hylan gel slurry can be degraded by all types of hyaluronidases. Tissue compartments such as the synovial space in joints and tendons, vitreus and subcutaneous intercellular space do not contain free hyaluronidases. Therefore hylan gel remains in these compartments unless it is broken up by physical forces (wear and tear). Microscopic pieces of hylan gel can enter the lymph system where they will be absorbed. In the vitreus of primates, we found the gel to be present after three years.

Q: J. Sunamoto (Kyoto Univ.)
(1) Can you expect the time independent release of drugs?
(2) What is the degradation rate in vivo ?

A: (1) I can expect only time-dependent release of drugs from hylan gels.
(2) Hylan fluid diffuses or flows away--depending on the movements of the tissue compartments -- and enters the blood circulation via the lymph system. Hylan gel stays in the tissue compartment unless physical forces break it up. Neither hylan fluid nor gel is degraded at the location where applied.

Q: K. Nishinari (Nat. Food Res. Inst. Japan)
The shock absorber must satisfy certain conditions. Can it be explained from the frequency dependence of elastic coefficient, for example, in case of traffic accident ? It also must have a peculiar large deformation behavior. Any comment on this ?

A: The frequency dependence of the dynamic elastic and viscous moduli shows that at high frequency (2-3 Hz) the human synovial fluid and hylan fluids have predominantly elastic properties. At low frequencies (0.01-0.05 Hz), these fluids show predominantly viscous properties. The rapid transition from predominantly viscous to predominantly elastic properties with increasing frequencies is characteristic for these polymers. When a person runs, the estimated frequency of joint loading is 2-3 Hz. One assumes that the impact on the joints in accidents produces even higher frequencies. Therefore our native shock-absorbing system, the hyaluronan molecular matrix, and hylans provide elastic protection under these conditions.

OLIGOSACCHARINS ACTIVATE DEFENSE RESPONSES IN PLANTS

P.Albersheim, C.Augur, P.Bucheli, F.Cervone, A.G.Darvill,
G.DeLorenzo, S.H.Doares, N.Doubrava, S.Eberhard,
D.J.Gollin, T.Gruber, V.Marfa-Riera and D.Mohnen
Complex Carbohydrate Research Center and Department of
Biochemistry, University of Georgia, Russell Research Center
P.O. Box 5677, Athens, GA 30613, USA

ABSTRACT[*]

Oligoglucoside fragments of a major structural poly-saccharide of fungal cell walls can elicit plant cells to accumulate phytoalexins (antibiotics), a plant response that appears to be a general defense mechanism against potential pathogens. The enzymes that catalyze the synthesis of phyto-alexins are themselves synthesized *de novo* in response to elicitors. An active hepta-β-glucoside and seven closely related but inactive hepta-β-glucosides were isolated from the mycelial walls of a fungal pathogen. These heptagluco-sides were structurally characterized and the active hepta-glucoside chemically synthesized. One ng of the synthetic or natural active heptaglucoside stimulates phytoalexin accumu-lation in soybean cotyledons.

Bacteria can elicit the accumulation of phytoalexins in plant tissues by secreting enzymes that release an oligo-galacturonide fragment of a plant cell wall polysaccharide that elicits the synthesis of phytoalexins. This "endogenous elicitor", the first-recognized biologically active carbo-hydrate from plant cell walls, was shown to be an α-1,4-dodeca-D-galacturonide.

The heptaglucoside elicitor from fungal cell walls acts synergistically with the oligogalacturonide elicitor from plant cell walls. When the heptaglucoside and oligogalacturo-nide elicitors are applied simultaneously to plant tissues, their combined elicitor activity is as much as 35-fold higher than the sum of the responses of these elicitors assayed separately. The results suggest that oligogalacturonides act as signals of tissue damage. This hypothesis is supported by reports that oligogalacturonide and glucan elicitors can activate other defense responses of plants, such as accumula-tion of lignin, hydroxyproline-rich glycoproteins, and pro-tease inhibitors.

Another mechanism by which plants resist microbial invasion is the hypersensitive response. This response is characterized by rapid death of plant cells at the point of attempted infection. An oligosaccharide fragment of cell walls of either suspension-cultured maize or sycamore cells, when applied to either of the two types of cells in liquid culture, causes cell death. We hypothesize that (1) the putative hypersensitive-inducing oligosaccharide is released by microbial enzymes during attempted infection of plants, and (2) the initiation of the hypersensitive resistance response by the released oligosaccharide is the key step in eliciting those responses of plants capable of defending them against potential pathogens.

More recently, we obtained evidence that oligosaccharins may regulate morphogenesis (Nature (1985)314:615). This evidence was obtained by incubating tobacco thin cell layers (TCLs) on liquid media in the absence or presence of plant cell wall fragments. We have optimized the TCL bioassay so that it is reproducible and have shown that pectic fragments can inhibit the formation of roots and induce the formation of vegetative shoots.

We call naturally occurring carbohydrates with biological regulatory functions "oligosaccharins".

(Supported by U.S. Dept.of Energy Grant #DE-FG09-85ER13425 and NIH Postdoctoral Fellowship #F32GM12372.)

* Only the abstract for the lecture of Prof. Albersheim is given here in accordance with his desire.

DISCUSSION

Q: G. Franz (Univ. Regensburg)

(1) How does the RGII fit into the Albersheim cell wall model ?

(2) Why were the KDO, DHA and acetic acid always overlooked when analyzing cell wall polysaccharides ?

(3) Is there any proof for the existence of oligosaccharide receptors ?

A: (1) We do not have a current model of the primary cell wall. Scientists are learning a great deal about the structures of primary cell wall polymers and are beginning to learn something about the distribution of the polymers within the primary cell walls but too little is known of the distribution and interconnections of wall polymers to design valuable model of the cell wall.

(2) Because they (KDD, DHA and acetic acid) are destroyed by the methods commonly used to analyze cell wall components.

(3) There is no firm evidence establishing the existence in plants of receptors for biologically active carbohydrates.

Q: R. M. Brown, Jr. (Univ. Texas)

(1) Role of oligosaccharins in practical applications in plant pathogenesis and agriculture ?

(2) Can oligosaccharins enter the cell ?

(3) Do oligosaccharins added to protoplasts influence cell morphogenesis, gene expression, etc. ?

A: (1) It is too early to delineate the role of carbohydrate regulatory molecules in agriculture but there are a lot of possibilities.

(2) We do not know whether oligosaccharins enter cells.

(3) No evidence has been obtained that oligosaccharins are able to influence morphogenesis or gene expression in protoplasts.

Q: T. Uryu (Univ. Tokyo)

Would you tell me about the structure of glycan portion of the glycoprotein (gp120) existing on the AIDS virus surface ?

Because my colleague, who is a professor of virology, said that the glycan portion may originate from cells, but not from the virus itself. Nobody knows the origin of the glycan portion.

A: The glycan chains of viruses are encoded by and synthesized by the cells of the host in which the virus multiplies. AIDS virus coat protein has about 25 N-linked carbohydrate chains, and perhaps five or ten different chains can be attached to each N-linked (asparagine) attachment site. It is a very complex problem to study this coat protein. There are two preliminary reports in the literature describing the complexity of this problem.

Q: K. Hatanaka (Univ. Tokyo)

In the case of oligogalacturonide, 10~13 mers are active. And the heptaglucoside is active. Then, more than octamer of oligoglucoside containing active heptaglucoside is still active ?

A: Yes, larger oligoglucosides that contain the active heptaglucoside are thought to be active. It may be that enzymes (β-glucanases and β-glucosidases) known to be present in the walls of plant cells convert the larger oligoglucosides into the active hepta-β-glucoside.

Q: H. Kuzuhara (Inst. Phys. Chem. Res. Japan)

Cell wall polysaccharide fragments of small molecule have potent activity to control physiological property. Do you think this type of phenomenon is general among organisms ?

A: We hypothesize that a variety of small oligosaccharides, released from cell walls by substrate-specific endoglycanases and further processed by specific glycosidases, are active as regulatory molecules in plants and animals.

POLYSACCHARIDES FOR PHARMACEUTICAL AND MEDICAL APPLICATIONS IN DRUG DELIVERY SYSTEMS

JUNZO SUNAMOTO
Laboratory of Materials Science of Polymers and Artificial Cell Technology
Department of Polymer Chemistry, Faculty of Engineering, Kyoto University,
Sakyo-ku, Yoshida Hommachi, Kyoto 606, JAPAN

ABSTRACT

Several naturally occurring polysaccharides, such as pullulan, amylopectin, and mannan, were chemically modified in part by conjugating cholesterol as the hydrophobic anchor and another saccharide determinant or fragment of monoclonal antibody as the sensory device for targeting a specific cell. By coating the outermost surface of liposome or O/W-emulsion with such polysaccharide derivatives, these colloidal particles were able to be employed as a cell specific drug carrier in the site specific drug delivery system. In this article, several successful results in the treatments of infectious and cancer diseases, in the enhanced macrophage activation, and in the serum free culture of fibroblasts are introduced.

INTRODUCTION

Needless to say, effective delivery of drugs to target cells or tissues brings about a decrease of toxic side-effects and an increase in the pharmacological activity of drugs, and subsequently makes a decrease in the dosage of drugs possible. This means that a drug delivery system (DDS) involves two important problems; namely, targeting and sustained release of drugs.

More precisely, "targeting" is classified into two different categories based upon the difference in its mechanism *in vivo* after administration (1). One is *passive targeting* with bulk recognition mechanism, which is attained by altering the bulk structural characteristics of the carrier such as the hydrophobicity (or lipophilicity), the charge density, the fluidity, the size, and/or the total morphology of the drug carrier. Therefore, this mechanism may be more important when the drug carrier and the target cell are apart from a long distance each other. It is now well known that passive targeting using small colloidal particles as drug carriers generally leads, after intravenous administration, to their uptake by the mononuclear phagocytes of the reticuloendotherial system predominantly in the liver and spleen and by circulating blood monocytes. Another and more specific

247

mechanism is *active targeting*, which is attained by molecular recognition
mechanism. A sensory device, such as an antibody, an antigen, a saccharide, or an
oligo peptide can be covalently conjugated or noncovalently attached to the drug
carriers.   In this mechanism, the targeting is attained by the recognition at the
more molecular level through the direct and specific interaction between the
recognition site of the drug carrier and the receptor of the target cell.   Therefore,
this mechanism becomes more effective when the carrier is very close to the
target cell in a very short distance (within van der Waals distance).

In the past two decades, the concept of targeting has been gradually
changed and the target itself has become more precise.   At the first generation,
"targeting" was to simply direct the drug toward a specific organ or tissue such as
the lungs, the kidney, the liver, and so forth.   At the second generation, however,
the target became the cellular level.   A specific cell of the organ or tissue under
the consideration is the real target, not the whole of organ or tissue.   In the third
generation we are probably going to think about the specific part of the target
cell, such as plasma membrane, cytosol, phagosome, mitochondria, nucleus, or
other small organs of the cell.   This means that we have to consider not only the
place to target but also the mechanism of the interaction between drug carrier
and cell, such as simple adhesion, phagocytosis, or fusion.   In order to establish a
truly targetable drug delivery system, therefore, we must carefully control the
molecular design of the drug carrier.

Liposome has been widely accepted to be a possible drug carrier, especially,
for water soluble drugs.   For lipophilic drugs, we can employ another better carrier such as O/W-emulsion.   These colloidal particles have been used *in vivo* as
drug carrier for passive targeting.

Since 1982, we have been extensively studying on the polysaccharide-coated
liposomes in order to make liposome more physically and biochemically stable and
more cell specific.   Recently, we have succeeded to apply this technique also to
O/W-emulsion besides liposome (Figure 1). In this paper, several recent results
which show the usefulness of these polysaccharide-coated particles in the treatment of several infectious diseases in animals, in the activity enhancement of
immunopotentiator, and in the serum free culture of fibroblasts will be introduced.

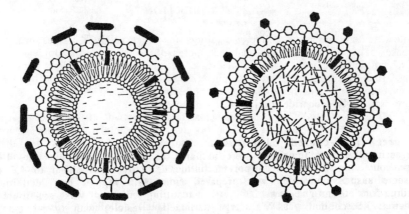

Figure 1.   Schematic representation of structure of polysaccharide-coated liposome
(left) and O/W-emulsion (right).

MATERIALS AND METHODS

Synthesis of polysaccharide derivatives.

Phosphorylation of mannan (M) was carried out by mixing mannan with polyphosphoric acid in the presence of formamide in DMSO. Sialic acid (N-acetyl neuraminic acid, NeuNA) was introduced to glucopyranose of a polysaccharide (pullulan (P) or amylopectin (Ap)) by glycosyl bond. Amino saccharide was introduced to pullulan by condensation with glucopyranose unit. Structures of polysaccharide derivatives are shown in Figure 2. Obtained compounds were abbreviated, for example, as NeuNA-3.8-CHP-50-1.9, which is pullulan (molecular weight, 50,000) substituted with 3.8 of NeuNA and 1.9 of cholesterol (CH) moieties per 100 glucose units.

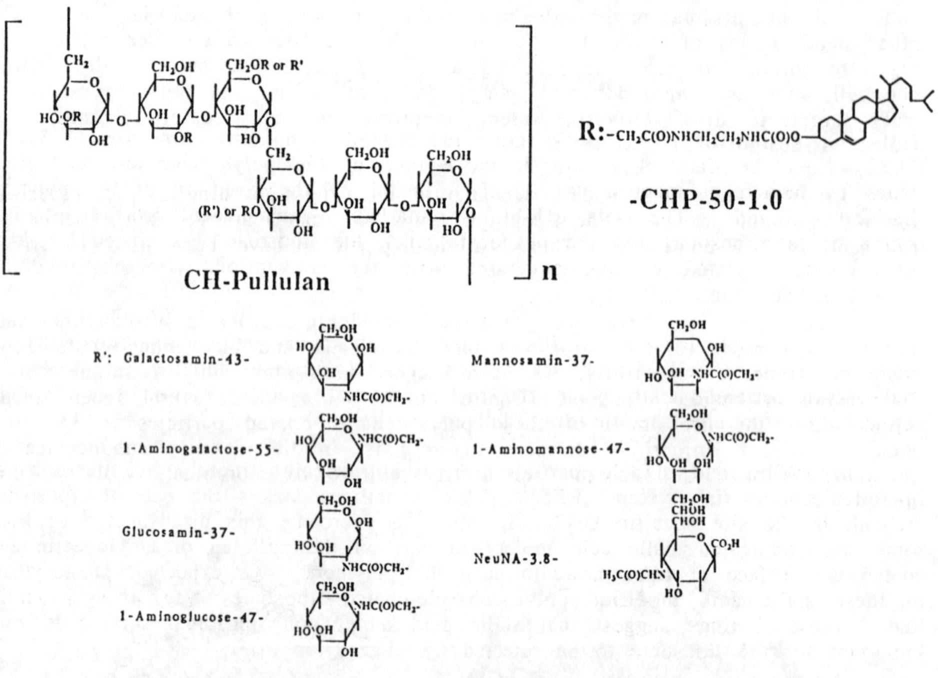

Figure 2. Polysaccharide derivatives employed.

Polysaccharide-coated liposomes.

Polysaccharide-coated liposomes were prepared as follows. For example, 1.0 ml of saline containing 7.5 mg of a polysaccharide derivative was added to 4.0 ml of a liposome suspension which was prepared from 37.5 mg of egg lecithin. After stirring for 30 min below 20°C, the resulting mixture was gel-filtered on a Sepharose 4B column. LUV's (large unilamellar vesicle) with or without polysaccharide derivatives were prepared by the reverse phase evaporation technique.

Polysaccharide-coated emulsions.
O/W-Emulsions were prepared with 100 mg of soy bean oil, 12 mg of egg lecithin, (consisting of 67.0 mol% phosphatidylcholine, 18.6 mol% phosphatidyl-ethanolamine, and 0.8 mol% lysolecithine), and 25 mg of glycerin. The mean diameter of the emulsion so obtained was 270 ± 60 nm determined by a Coulter N4 Submicron Particle Analyzer. Polysaccharide-coating of the emulsion was performed by briefly sonicating the mixture of the emulsion and an aqueous solution of a polysaccharide derivative.

Measurements.
Lectin-induced aggregation, cell uptake *in vitro*, and tissue distribution of polysaccharide-coated liposomes and emulsions were investigated by the same methods described in literatures(2,3). For the emulsions, microelectrophoretic measurements were carried out on a Laser Zee Meter Model 501 (Penkem, Inc., New York).

RESULTS AND DISCUSSIONS

Cell recognizability of polysaccharide-coated liposome *in vitro* (2,3).
Clearly from the data' of Figure 3, the uptake of the polysaccharide-coated liposomes by human monocytes was largely affected by the structure of the polysaccharide employed. The data exhibit a remarkable enhancement in the internalization of the liposome into the phagocyte when the liposome was coated by mannan or phoshorylated mannan compared with the conventional liposome without polysaccharide coat.

Figure 4 shows the specific uptake of phosphorylated mannan-coated liposome by mouse L-cell (fibroblast). These results suggest that the receptor-mediated uptake certainly takes place and that these cell specific liposomes are widely applicable to the site specific drug delivery system.

Lipophilic and sizable particle are usually cleared rapidly by macrophages in reticuloendotherial system (RES). This sometimes makes the use of liposomes difficult in the site specific DDS. In order to overcome this disadvantage of liposome, we conjugated sialic acid moiety in part on the pullulan or amylopectin and coated the surface of liposome with such the polymers. As expected, phagocytosis of these sialic acid modified polysaccharide-coated liposomes was strongly inhibited. These findings suggest that sialic acid moiety on liposomal surface is very important for the liposome to be rejected by phagocytes (4).

In vivo **study of polysaccharide-coated liposomes.**
Tissue distribution and blood clearance of the polysaccharide-coated liposome were investigated. Mannan or amylopectin-coated liposomes were highly accumulated in the lungs after intravenous administration. This was consistent with the results in the macrophage uptake of these modified liposomes (5). On the other hand, when liposome was coated by sialic acid-conjugated polysaccharide and intravenously injected into guinea pigs, the clearance rate of NeuNA-3.8-CHP-50-1.9-coated liposome from blood stream decreased two times compared with that of non-coated liposome. And, tissue distribution was also altered depending on the structure of polysaccharide. Mannan-coated liposome showed an increase in the lung uptake, while in the case of sialic acid-modified amylopectin-coated liposome, the lung uptake was significantly depressed. When galactosamin-41-CHP-1.9-coated liposome was intravenously injected into C57BL/6 mice, liver uptake increased at 30 min after i.v. injection compared with the simple CHP-50-1.9-coated

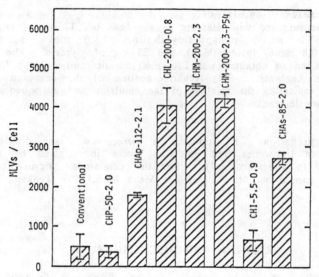

Figure 3. Internalization efficiency of various liposomes labelled with [14C]-DPPC into human monocyte at 37°C for 30 min in RPMI-1640: [MLV]/[Cell] = 6.6×10^5

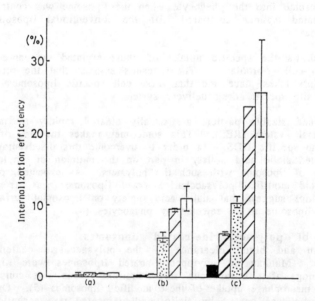

Figure 4. Internalization efficiency of various liposomes labelled with [14C]-DDPC into L-Cell at 37°C for 1 hr(■), 2 hr(▨), 6 hr(▨),12 hr(▨), and 24 hr(☐) in SFM-101: [MLV]/[Cell] = 1.25×10^6; (a) non-coated liposome, (b) CHM-coated liposome, and (c) CHM-P-(4)-coated liposome.

liposome. These findings suggest to us that the terminal sugar moiety of modified polysaccharides behaves as an excellent sensory device of drug carrier in site specific DDS.

Colloidal stability of polysaccharide-coated emulsion.
In the absence of polymers a relatively rapid increase in the turbidity was observed. However, if either CHAp(cholesterol-bearing amylopectin)-112-1.8, CHP(cholesterol-bearing pullulan)-50-1.9, EHEC (ethyl (hydroxyethyl)-cellulose)-146 or pullulan-50 was added to the emulsion before the addition of Ca^{2+}-ion, the Ca^{2+}-induced aggregation was drastically depressed (6). In addition, when the emulsion was coated by amylopectin derivative, both the macrophage uptake *in vitro* and the lung uptake *in vivo* were enhanced.

Application of polysaccharide-coated liposomes in chemotherapy and immunotherapy.
The best treatment of bacterial infectious diseases involves the use of antibiotics. In 1983, we tried to treat experimental *Legionnaires'* disease in guinea pigs employing a macrophage specific liposomal carrier (7). In this treatment, sisomycin was encapsulated in OPA-112-(4.9)-coated LUV. When the disease was treated with free antibiotics without liposomes, all the guinea pigs died within 6 days, while 100 % survival rate was attained when the animals were treated with a mixture of free (64 %) and liposome-encapsulated antibiotics (36 %). To our knowledge, this was the first successful demonstration of the treatment of bacterial infectious disease with an antibiotics-bearing and cell specific liposomes. *Pulmonary Candidiasis* also was treated with amylopectin-coated liposome containing amphotericin B (AmpB) (8). The LD_{50} of AmpB is 1.2 mg/kg, but when amylopectin-coated liposome bearing AmpB was administered all mice survived even at the dose of 5 mg/kg. This indicates a remarkable reduction in toxicity of AmpB by employing the polysaccharide-coated liposome delivery system. In addition, the amylopectin-coated liposome showed remarkable efficacy in the survival rate (100 %), which was much better than that of the conventional liposome (50 %) even at the same dosage.

One difficulty in utilization of liposomes as drug carriers is that liposomes are cleared rapidly by macrophages in the RES. However, delivery of drugs such as immunomodulator to macrophages is not a disadvantage. Activated macrophages may participate in the killing of bacteria of tumor cells. In order to modulate the *in vivo* adjuvant activity of an immunopotentiator, encapsulation of an immunopotentiator, such as synthetic polyanionic polymers or artificial DNA analogues, into liposomes has been attempted, and successful results were obtained (9,10).

Treatment of tumor bearing mice using newly developed immunoliposome (11-13).
We have recently developed a new and improved technique for safely binding a monoclonal antibody fragment onto the liposomal surface. This method involves coating of the outer surface of large liposomes with the polysaccharide derivative and subsequent binding of SH bearing antibody fragment (IgMs or Feb') to the pullulan derivative on the liposomal surface. Using such new immunoliposome loading adriamycin, lung tumor-bearing mice were treated. In the case of immunoliposome, a statistically significant inhibition of tumor growth (P<0.005) was observed and the tumor was even regressed completely at 2 weeks after the administration.

Serum free culturing of fibroblasts in the presence of cell specific liposome (14).

Mammalian cell culture in serum free medium supplemented with well defined materials is very important in the progress of biotechnology for producing bioactive compounds. We have also developed a new technique with the use of egg lecithin liposomes as coated by phosphorylated mannan derivative instead of serum as the carrier of water-insoluble essentials to cells. The growth ratio of mouse L-cells certainly increased in the presence of serum compared with the case of serum free. However, when phosphorylated mannan-coated liposomes were employed instead of serum, the growth ratio became apparently larger than the case with serum.

CONCLUSIONS

Liposome and O/W-emulsion are relatively unstable both mechanically and chemically and less cell specific. In order to utilize these colloidal particles in the wider field of biotechnology, we have to remove these disadvantages. By coating the surface of these particles with naturally occurring polysaccharide derivatives, we could succeed in making these particles more stable against external stimuli and more cell recognizable as well. Such the modified colloidal particles were applicable in various aspects of biotechnology such as medicine and cell culture.

REFERENCES

1. Sunamoto, J., and Iwamoto, K., Protein-coated and polysaccharide-coated liposomes as drug carriers, The CRC Critical Reviews in Therapeutic Drug Carrier systems, 1986, 2, 117-136.

2. Sunamoto, J., Iwamoto, K., Takada, M., Yuzuriha, T. and Katayama, K., "Improved drug delivery to target specific organs using liposomes as coated with polysaccharides" in: Polymers in Medicine", Chiellini, E. and Guisti P., (Eds), Plenum Publishing Co., 1984, pp. 157-168.

3. Sunamoto, J., Iwamoto, K., Takada, M., Yuzuriha, T. and Katayama, K., "Polymer coated liposomes for drug delivery to target specific organs", in: "Recent Advances in Drug Delivery Systems", Anderson, J. M. and Kim, S. W., (Eds.), Plenum Publishing Co., 1984, pp. 153-162.

4. Sunamoto J., Sakai, K., Sato, T. and Kondo, H., Molecular recognition of polysaccharide-coated liposomes. Importance of sialic acid moiety on liposomal surface, Chem. Lett., 1988, 1781-1784.

5. Takada, M., Yuzuriha, T., Katayama, K., Iwamoto, K. and Sunamoto, J., Increased lung uptake of liposomes coated with polysaccharides, Biochim. Biophys. Acta, 1984, 802, 237-244.

6. Carlsson, A., Sato, T., and Sunamoto, J., Physicochemical stabilization of lipid microspheres by coating with polysaccharide derivatives, Bull. Chem. Soc. Japan, 1989, 62, 791-796.

7. Sunamoto, J., Goto, M., Ihda, T., Hara, K., Saito, A. and Tomonaga, A., "Unexpected tissue distribution of liposomes coated with amylopectin derivatives and successful use in the treatment of experimental Legionnaires' diseases", in: "Receptor-Mediated Targeting of Drugs", Gregoriadis, G., Poste, G., Senior, J. and Trouet, A., (Eds.), Plenum Publishing Co., 1985, pp. 359-371.

8. Kohno, S., Miyazaki, T., Yamaguchi, K., Tanaka, H., Hayashi, T., Hirota, M., Saito, A., Hara, K., Sato, T. and Sunamoto, J., Polysaccharide-coated liposome with antimicrobial agents against intracytoplasmic pathogen and fungus, *J. Bioactive and Compatible Polym.*, 1988, **3**, 137-147.

9. Sato, T., Kojima, K., Ihda, T., Sunamoto, J. and Ottenbrite, R. M., Macrophage activation by poly(maleic acid-alt-2-cyclohexyl-1,3-dioxap-5-ene) encapsulated in polysaccharide-coated liposomes, *J. Bioactive and Compatible Polym.*, 1986, **1**, 448-460.

10. Ottenbrite, R. M., Sunamoto, J., Sato, K., Kojima, K., Sahara, K., Hara, K. and Oka, M., Improvement of immunopotentiator activity of polyanionic polymers by encapsulation into polysaccharide-coated liposome, *J. Bioactive and Compatible Polym.*, 1988, **3**, 184-190.

11. Sunamoto J., Sato, T., Hirota, M., Fukushima, K., Hiratani, K. and Hara, K., A newly developed immunoliposome — an egg phosphatidylcholine liposome coated with pullulan bearing both a cholesterol moiety and an IgMs fragment —, *Biochim. Biophys. Acta*, 1987, **898**, 323-330.

12. Sato, T., Sunamoto, J., Ishii, N. and Koji, T., Polysaccharide-coated immuno-liposome bearing anti-CEA Fab' fragment and its internalization into CEA-producing tumor cells, *J. Bioactive and Compatible Polym.*, 1988, **3**, 195-204.

13. Hirota, M., Fukushima, K., Hiratani, K., Kadota, J., Kawano, K., Oka, M., Tomonaga, A., Hara, K. Sato, T. and Sunamoto, J., Targeting cancer therapy in mice by use of newly developed immunoliposomes bearing adriamycin, *J. Liposome Res.*, 1988-89, **1**, 15-55.

14. Sunamoto, J. and Sato, T., Serum free culture of mammalian cells on a bio-polymer in the presence of cell specific polysaccharide-coated liposomes, *Polymer Preprints, Japan*, 1987, **36**, 2730.

DISCUSSION

Q: G. Franz (Univ. Regensburg)
What are the optimal polysaccharides to be utilized for coating liposomes ? Linear, branched, or chain length (DP) ?

A: In order to describe the optimal condition to coat liposomes with a polysaccharide, we have to think about two aspects : (1) the coating efficiency for simply making liposome more mechanically stable, and (2) for obtaining the better cell specificity.

(1) In order to obtain the more stable liposome, the linear (in solution) structure of the polysaccharide is desirable. The larger molecular weight and the less branched polymers are the better. In addition, the substitution degree of the hydrophobic anchor, cholesterol moiety, is hopefully 1 or 2 per 100 monosaccharide units. From this point of view, cholesterol-bearing pullulan seems the best at present.

(2) If you say about the cell specificity of the polysaccharide-coated liposomes, relatively branched polysaccharide is better, since the terminal saccharide moiety must be somewhat flexible and extended to the bulk aqueous phase from the liposomal surface. If this is not the case, the receptor on the target cell surface can not recognize and bind effectively the saccharide moiety on the surface of the liposome. In this sense, amylopectin or mannan was better in our studies.

As the result, pullulan, which is bearing an appropriate number of monosaccharide moiety that is conjugated with a suitable spacer and can play a role of the cell recognition site, is considered to be the optimal polysaccharide.

SUMMARY

given by C. Schuerch for Round Table II-2

The subject areas of the speaker's talks were explored in more depth during the Round Table. No specific conclusions were reached but there was general agreement that these were outstanding paper, on a most important emerging frontier of science.

SPECIAL CONTRIBUTIONS

ADVANCES IN THE CHEMISTRY OF WOOD AND CELLULOSE IN ROMANIA

CRISTOFOR I. SIMIONESCU and VALENTIN I. POPA
Polytechnic Institute of Jassy, Romania

ABSTRACT

In recent years, Romanian chemistry of wood and cellulose has witnessed significant progresses, referring to the diversification and extension of some theoretical and practical aspects. Starting from this observation, the present paper analyzes the results of the efforts of assuring the necessary resources of raw materials, as well as of the study and processing of their components in pulp and paper industry. There are also discussed the latest findings in the process of turning to good account some byproducts (secondary resources) resulted from wood processing industry or from agricultural and energetical cultures.

INTRODUCTION

Wood and cellulose chemistry has been scarcely represented in pre-war Romania, both from the industrial point of view and from that of developing modern techniques and training specialized staff. Starting with 1944, to the modest small 12 factories, 7 big pulp and paper mills have been added, so that a 20-fold increase of pulp and, respectively, a 13-fold increase of paper industry - as compared with 1938 - has been recorded.

In parallels, important steps have been taken in the action of training specialized staff; in this respect, the Department of Pulp, Paper & Man-made Fibers has been created in 1948, within the Polytechnic Institute of Jassy.

In the years to come, education in this field has been backed up by constant preoccupations for the development of research fields that should fully answer to the necessities of the rapidly growing industry. In several new areas of investigation, Romanian researchers have acquired genuine results, a fact which explains the top position occupied by Romania in the international competition within the field of pulp, paper and rayon fibers scientific investigation.

Two journals are being issued in Romania on this topic: Celuloză și Hîrtie (since 1951) and Cellulose Chemistry and Technology (since 1967); also, 9 international symposia on cellulose chemistry and technology have been held in Jassy, starting with 1961.

The investigations undertaken - out of which we shall only mention those of the last decade - cover large themes, including - among others - problems peculiar to the field of pulp and paper, such as providing, and exploiting, too, the existent resources, on having permanently in view the problem of assuring environmental preservation and protection.

1. Raw material resources

One of the main characteristics of Romanian forest fund refers to the prevalence of hardwood species, beech wood especially, a fact which, as a consequence, defines the orientations of wood processing industry. At the same time, the concern for covering the necessary raw materials in pulp and paper industry induced preoccupations for diversifying the existing resources, by increasing mainly the utilizations of hardwood species (poplar wood, especially) and annual plants (straw and reed, as well).

In the process of valuation, characterization and turning to good account of such resources, several complex stages have been run through. Thus, in the last 25 years, the main efforts of the Romanian specialists have been directed towards accumulating and providing a rich experimental material, both of theoretical and practical interest, synthesized in three consistent monograph studies (1-3). The experience gained has been further developed by deepening the studies for the characterization of anatomo-morphological elements of poplar and willow wood, alongwith the concern for finding the ways of transforming them into paper products. Such peculiarities involve special technological efforts of practical application and industrialization; more than that, the structural complexity of the raw materials and of the products to be manufactured has required some statistical correlations between the chemical-biometrical properties of wood and the physico-mechanical indices of the obtained pulps (4).

In view of preventing the shortage of woody materials, Romania has to follow, in the coming decades, a rational policy of preservation, exploitation and accumulation of forest resources. In this respect, the necessary measures to be taken refer to: i) intensification of the action of recovering and substituting weakly-productive tree species; ii) increase of the forest surfaces; iii) extension of cultures of softwood and fast-growing hardwood and softwood species; iv) utilization - in cultures - of selected trees; v) application of fertilizers, alongwith irrigations and drainings; vi) setting up of Euramerican poplar wood; vii) afforestation of some reedy and piscicultural soils, as well as of agricultural degraded ones.

In parallels, turning to good account of some annual plants, of certain energetic cultures, alongwith those of recycling fibrous byproducts, will be performed.

2. Chemistry of Wood and its Components

Starting from the richness of the existent vegetal resources,
studies of the characterization of poplar hemicelluloses have
been developed, in parallels with those of applying quantum-
chemical methods in the analysis of lignin and of determining the chemical composition of poplar and willow wood (5),
and with the analysis of the thermal behaviour of certain
compounds, subjected to various processings.

Preoccupations for the complex turning to good account of
biomass have been carried out in studies discussing the methods of separation and characterization of photosynthesis-
ing pigments and also, of some secondary components extracted
through various procedures.

Investigation of cellulose's structure, by various methods, has rendered evident the differences appearing at
supermolecular level, in the case of cellulosic products of
various origins, that have been subjected to physico-chemical treatments.

Diversification of the fields of cellulose application
has been the theme of various investigations; thus, mention
must be made of the contributions discussing the establishment of the optimal conditions for the fabrication of micro-
nized celluloses, as well as of some correlations between the
existent technologies and the field of the products' application (6). Materials possessing new properties have been synthesized by grafting secondary amines on fabrics from fibrous
mixtures (aiming at improving the dyeing capacity), by grafting acrylamide and acrylonitrile on straw pulp (in view of
obtaining products with paper properties), grafting of rayon
fibres with phosphorous polymers and, finally, grafting of
lignin with vinylic monomers.
Concomitantly with the application of the classical procedures (7) of initiating grafting reactions, new methods have
been developed, out of which the most promising one was
found to be the method utilizing cold plasme conditions (8).

Cellulosic fibres, modified under different conditions,
have been characterized through classical physico-chemical
techniques, and also by the method of establishing the elec-
trokinetic potential and that of discharging thermally stimulated currents.

Other studies were dedicated to the synthesis of cellulose derivatives possessing reactive groups (9) and of other
natural polymers that may be advantageously utilized in the
immobilization of some enzymes, antibiotics or vitamins.

In view of synthesizing the information necessary for the
application in practice of some laboratory investigations, a
series of aspects - referring to the synthesis of carboxy-
methylcellulose and cellulose acetate - have been analyzed;
certain cellulosic and lignin products have been characterized from the viewpoint of their polydispersity.

At present, worth mentioning are the concerns for the
application of some special procedures of textile finishing
of cotton fabrics, of obtaining some polyfunctional cellulosic derivatives, and of producing fibres from natural and
synthetic polymer mixtures.

3. Pulp Technology

In the field of pulp technology, investigations have been performed on the utilization of natrium and ammonium bases (10) in processes of sulphite delignification. At the same time, there have to be mentioned the contributions brought to the optimization of the technological stages involved in the fabrication of pulps to be used in chemical processings; thus, analyzes have been employed, both on the process of sulphite delignification and on that of prehydrolysis sulphate (for hard- and softwood species); also, there have been establish-ed the conditions of alkaline extraction and those of bleaching with conventional agents or with molecular oxygen, too.

The great variety of the fibrous raw materials applied in pulp and paper industry induced the initiation of some investigations upon the possibilities of processing straw, by applying the NSS method, and juvenile poplar wood, by the sulphate procedure (5). In recent years, significant results have been obtained in the application of additives such as anthraquinone in sulphate or natron cooking of hardwoods or rice straw, concomitantly with the preparation of some agents, showing similar functions, through residual lignin processing. At the same time, trials have been made for the application of certain oxido-alkaline processes of softwood delignification.

Actual investigations are mainly oriented towards the optimization of the classical procedures of paper and dissolving pulps fabrication, the introduction of some additives with catalytic functions and the application of processes based on the utilization of some non-conventional delignification agents, such as organic solvents.

4. Paper Technology

In this field of investigation, there have been further developed the preoccupations for the fabrication of papers, both from man-made fibres and from their mixtures with cellulosic fibres; thus, a series of special, filtering papers have been obtained from polyacrylonitrile fibers (their behaviour being followed in the process of beating) (11), as well as wall papers, concomitantly with the determination of the behaviour of some paper products during their drying.

In parallels, investigations have been initiated for setting precisely the technological conditions to be met in the fabrication of some technical paper grades, by using various native fibrous raw materials, characterized by the above-mentioned peculiarities.

The newly-appeared conditions of saving material and energetic resources brought into activity a series of earlier aspects, referring to the fabrication of paper in neutral or weakly-alkaline medium. Thus, there have been evaluated the possibilities of using natrium aluminate, activated with hemicellulose, in the fabrication of paper (12). At the same time, there have been studied the conditions of utilizing calcium carbonate as filler for paper, both in classical sizing conditions and in the presence of alkyldimerketenes (13).

Another series of investigations deal with treatments of paper covering and with the fabrication and utilization of new macromolecular additives, to be used in the improvement of various paper products characteristics. A special importance has been given to increasing the degree of waste papers utilizations, as well as of some secondary fibrous resources resulted from the fabrication and processing of pulp products.

5. Secondary Resources of Raw Materials and Energy

Special attention has been given, in recent years, to increasing the utilization degree of the raw materials from pulp industry, both by increasing the production yields and by applying new solutions, inducing a thorough valorification of all byproducts. In this respect, our own experience in the cultivation of fodder yeasts on waste liquors from sulphite delignification has led to a better knowledge of new aspects of this process and also of the biochemical processing of the hydrolysis products resulted from the fabrication of dissolving pulp. Certain requests for furfural, of the chemical industry, initiated investigations upon the most suitable raw materials and control methods to be followed in the establishment of the fabrication technology.

At the same time, the huge amounts of lignocellulose resulted from the fabrication induced new studies for finding out possible chemical utilizations of this byproduct. Thus, interesting results have been obtained by applying some chlorination, nitration and nitrooxidation reactions, which led to products further used as additives in the process of wood delignification, or as intermediates in the preparation of dyes (14, 15).

Recently, investigations dealing with rendering profitable waste lignins, through oxyammonolysis reactions, for the preparation of agricultural fertilizers, have been intensified. Also, mention must be made of the contributions to the synthesis of lignin phenolic and lignin epoxy resins, aiming at substituting phenol by lignin products (16).

The new conditions brought about by the crisis of raw materials and energy have stimulated new researches of phytomass.Thus, alongwith the determination of some accessible sources from the industry of chemical processing of wood, there have also been examined the possibilities of cultivation and processing of some vegetal resources with a high hydrocarbon contents (17). At the same time, on having in view the complex composition of vegetal biomass, a new procedure has been developed, permitting the sequential and modulated treatment of any type of phytomass (18).

Significant advances have been recorded in the investigation of the alkaline extraction stage, permitting the recovery of an important amount of hemicelluloses and polyphenols, products with large applications, among which those referring to the obtainment of some adhesive systems for wood processing industry (19) are the first to be considered. Lignocellulose resulted from alkaline treatment is a suitable support for the enzymatic and/or biological degradation (20).

At present, investigations are being developed for the diversification of the directions of utilization of waste lignins (by applying some substitution and condensation reactions), of polyphenolic products (for plant growth biostimulators) and of the enzymatic and biological treatment of lignocelluloses, for the production of proteinaceous products and of biochemical and mechanical pulps having paper characteristics.

Referring to this last aspect, there have been obtained - concomitantly with the action of selecting microorganisms - a series of preliminary results upon the control and regulation of the enzymes involved in the process of biodegradation of the vegetal biomass components.

Finally, in such a bewildering and, at the same time, highly promising field as that of cellulose chemistry and technology, the concerns of Romanian specialists aim mainly at approaching, in a new manner, original methods of supplying the necessary resources, and of extending their field, new, non-conventional technologies, as well as the problem of energy and recycling products.

REFERENCES

1. Simionescu, Cr.I., Grigoraş, M. and Asandei-Cernătescu,A., Chemistry of Wood from Romania, Ed. Academiei, Bucureşti, 1984.

2. Simionescu, Cr.I. and Rozmarin, G., Reed Chemistry, Ed. Tehnică, Bucureşti, 1966.

3. Simionescu, Cr.I., Rozmarin, G., Grigoraş M. and Asandei-Cernătescu,A., Wood Chemistry: Poplar and Willow, Ed. Academiei, Bucureşti, 1973.

4. Rozmarin, G., Aly, M.I.H. and Simionescu, Cr.I., Possibilities of turning to good account juvenile poplar wood in the paper and hydrolysis industry. III. Statistic correlations between chemico-biometric properties of wood and physical properties of kraft pulps. Celuloză şi Hîrtie, 1983, 32, 46-52.

5. Simionescu, Cr.I., Aly, M.I.H., Toma, C. and Rozmarin, G., Possibilities of turning to good account juvenile poplar wood in the paper and hydrolysis industry. I. Biometrical data and chemical composition. Cellulose Chem. Technol., 1984, 18, 55-62.

6. Oprea-Vasiliu, Cl., Negulianu, C., Popa, M., Weiner, F. and Nicoleanu, J., Contribution to the characterization of certain microcrystalline celluloses. Cellulose Chem. Technol., 1980, 14, 655-633.

7. Oprea, S., Dumitriu, S. and Bulacovschi, V., Grafting of acrylamide and acrylonitrile on some celluloses. Cellulose Chem. Technol., 1979, 13, 3-11.

8. Simionescu, Cr.I. and Denes, F., The use of plasma-chemistry in the field of the synthesis and modification of the natural macromolecular compounds. Cellulose Chem. Technol., 1980, 14, 285-316.

9. Simionescu, Cr.I., Dumitriu, S. and Bouzaher, Y., Bio-active polymers. IX. Cellulose derivatives with reactive groups. Cellulose Chem. Technol., 1982, 16, 67-75.

10. Gavrilescu, D., Obrocea, P. and Diaconescu, V., On wood delignification by sulphite process. Celuloză şi Hîrtie, 1983, 32, 176-190; 1987, 36, 99-101, 163-165.

11. Diaconescu, V., Popa-Stoicescu, M., Stoleriu, A. and Airinei, A. Synthetic papers from chemical fibres. II. Modifications of the polyacrylonitrile fibers in the beating process. Cellulose Chem. Technol., 1982, 16, 237-242.

12. Poppel, E. and Petrovici, V., Influence of sodium alu-minate with addition of hemicelluloses on the heating ability of some pulps and physico-chemical properties of neutral sized papers. Cellulose Chem. Technol., 1981, 15, 355-371.

13. Poppel, E. and Bobu, E., On the mechanism of paper sizing with alkyl diketene. Cellulose Chem. Technol., 1985, 19, 707-728.

14. Rozmarin, G. and Popa, V.I., Investigations in the field of lignin valorification. IV. Characterization of ligno-cellulose nitration products. Cellulose Chem. Technol., 1985, 19, 549-556.

15. Popa, V.I., Investigations in the field of lignin valo-rification. V. Products of oxidative nitration. Cellulose Chem. Technol., 1985, 19, 657-661.

16. Simionescu, Cr.I., Cazacu, G. and Macoveanu, M.M., Lignin epoxy resins. Physical and chemical characterization. Cellulose Chem. Technol., 1987, 21, 525-534.

17. Simionescu, Cr.I., Rusan, V., Roşu, D. and Caşcaval, N.C. Complex processing of vegetable biomass. Hydrocarbons from latex-producing plants. Cellulose Chem. Technol., 1984, 18, 343-348; 1985, 19, 357-363; 1987, 21, 113-120.

18. Simionescu, Cr.I., Rusan, V. and Popa, V.I., Options concerning phytomass valorification. Cellulose Chem. Technol., 1987, 21, 3-16.

19. Simionescu, Cr.I., Rusan, V., Cazacu, G., Macoveanu, M.M., Popa, V.I., Popa, M. and Bulacovschi, J., Polyphenol epoxy-AS resins based on alkaline extracts from the latex-bearing plant Asclepias syriaca L. I. Influence of the synthesis conditions. Cellulose Chem. Technol., 1987, 21, 639-649.

20. Simionescu, Cr.I., Popa, V.I., Rusan, M., Viţalariu, C. and Rusan, V. Biodegradation of vegetal materials under the action of some mixed cultures of microorganisms. Cellulose Chem. Technol., 1988, 22, 293-301.

Recent Developments in the Swedish Cellulose Industry - A Review

Bengt Rånby, Department of Polymer Technology, The Royal Institute of Technology, Stockholm, Sweden.

The industry based on wood as raw material is now called <u>forest industry</u> in Sweden. It comprises all activities of planting, growing and cutting trees in the forest, the use of wood as a construction material, the processing of wood into fiber material for paper, board, laminates and adsorbents and for chemical conversion into man-made fibers, cellulose derivatives, etc. Some wood and wood products are burned as fuel for heating of buildings, water, etc. In this review only advances related to the pulp and paper industry, here called <u>cellulose industry</u>, will be presented and discussed.

1. <u>Efficient Use of the Wood Raw Material from Softwood and Hardwood</u>

Sweden has large resources of forest land covering some 60% of the total area, i.e. about 27 million hectare. It is estimated that the annual growth on the forest land in Sweden is totally 90 million "forest m^3". This figure includes the whole volume of the trees above ground with bark and top included. The Swedish cellulose industry is presently using 32 million "wood m^3" (without bark and top) from the Swedish forests, 14 "wood m^3" is used for the saw mills and 10 "wood m^3" is burned. (1 "wood m^3" = 1.2 "forest m^3"). There is a shortage of wood for industrial use and 7 to 8 million "wood m^3" is imported to Sweden, mainly from the other Scandinavian countries, Soviet Union, West Germany and Chile. Estimating all logging of trees in Sweden, a total of 75 million "forest m^3" is accounted for.

This means that the cutting could be increased by some 15 million m^3 to balance the estimated annual growth. More efficient logging would make import of wood unnecessary.

To make logging more economical and efficient and increase the domestic supply of raw material for the cellulose industry, especially from privately owned forest land, new logging machinery is being developed. One mechanized logger, "MB-trac", is cutting, barking and chipping the trees in the forest with only small damage and loss of wood material. This powered machine is operated by two men and has a capacity of 6 m^3 per hour for trees of at least 10 cm diameter. Another small powered machine called "steel-horse" operated by one man, can pull about 1 m^3 of logs in forest terrain to logging roads where the logs are chipped or carried away by truck or tractor.

Hardwood was previously used mainly as fuel. Now there is an increased use of hardwood for pulping in the Swedish industry. About 15-17% of the forest trees in Sweden are different hardwoods, mainly birch. Because birch grows faster than softwoods (pine and spruce) the supply of hardwood is larger than the number of trees indicates. Pulp fibers from birch are shorter and slimmer than softwood fibers. They give therefore a denser and smoother paper but usually a paper with lower tear strength.

Previously it was difficult to use large amounts of hardwood fibers in papermaking. The paper machines have now been developed to run at high speed also with short hardwood fibers in large amounts. The technology is so successful that there is now a shortage of hardwood fibers for papermaking.

To increase the supply treebreeding projects have started to develop fast-growing hardwoods for the Swedish climate. These are hybrids of birch (Betula species) and aspen (Populus species). The pulps produced have advantages also in processing and use: they are easy to bleach without elemental chlorine, they don't turn yellow when stored and they give a soft paper, e.g. as tissue. Also high yield pulps of CTMP-type (Chemical Thermo-Mechanical Pulps) are now produced from aspen wood.

2. Developments in Pulping and Bleaching

The sulphate (or Kraft) process is a Swedish invention. It was originally a batch process, but it has been developed to a continuous process by (KAMYR AB). Presently the original batch Kraft process is being modified by Sunds Defibrator in Sweden and Beloit-Rader in USA to the Rapid Displacement Heating process (RDH). The principle is that the chips are initially impregnated with warm black liquor before the hot sulphate liquor is added. At a certain lignin content (measured as kappa-number) the RDH process has lower energy consumption, shorter process cycle, better washing and higher viscosity of the pulp cellulose than conventional Kraft.

The most important advances are presently being made in the bleaching of the pulps. Conventional bleaching is made by elemental chlorine, washing with alkali and usually a final step with chlorine dioxide. The chlorinated lignin compounds are biologically rather stable due to the chlorine/carbon bond, they cannot easily be re-covered or decomposed and are, therefore, bad water pollutants. Great efforts have been made to decrease the use of chlorine in bleaching of pulps by replacing it with oxygen and peroxide as bleaching agents. These efforts have been quite successful and the use of chlorine has been decreased to a fraction of previous use.

Experiments with new and unconventional bleaching agents are also in progress. In one process (PRENOX) invented by Olof Samuelson, the pulp fibers are pretreated with NO_2 vapor at low temperature (~40°C), then slurried in an acid nitrate solution which removes most of the lignin. In a final step the fibers are bleached with oxygen as in previously developed processes. The PRENOX-process has not as yet been used in full-scale operations.

Other bleaching processes under development are using oxidation-reducing sequences and washing with dilute alkali. Attempts are also made to eliminate residual chlorine when chlorine dioxide is produced.

3. Developments in the Paper Making Process

The economy in conventional papermaking by the wet process is largely dependant on the speed of the paper machine and the energy consumption in the removal of water. Only the second item will be discussed here.

When the wet fiber web on the wire net enters the forming and dewatering section of the paper machine, the pressing and suction-off of the water is made in an extended multiple nip of several elements which are quite efficient. This saves energy, because less water has to be removed by drying in this modified process.

A creative and promising development is the "impuls drying" method invented by Douglas Wahren some ten years ago and now appearing ready for full-scale application. This method is based on an ingenious thought. The wet sheet on the wire net is pressed against a very hot calander surface (~175°C) which is pressing and steam-blowing the water out of the fiber web and through the wire net.

Very rapid dewatering rates have been reached successfully for thin porous paper. Also thicker paper, e.g. liner and fluting for corrugated board, can now be produced by "impuls drying". The paper has higher density than conventionally dried sheets. This gives higher tensile, burst and compression strength, but the stiffness decreases. A serious problem is bubble formation and delamination of the sheet which may occur for thick paper qualities.

4. The Dioxin Problem

In 1985 it was reported from the USA that small amounts of very toxic dioxin compounds were found in the effluent of wash water and fiber sludge from pulp and paper mills. Careful analyses, especially in the USA and the Scandinavian countries have shown that very small amounts of several toxic substances are formed during the bleaching of pulp fibers with elemental chlorine and treatment with aqueous alkali. The main compounds found are TetraChloroDibenzoDioxin (TCDD) and TetraChloroDibenzoFuran (TCDF) and they are classed as strongly carciogenic by the Environmental Protection Agency (EPA) in the USA. The TCDD and TCDF amounts formed during bleaching of the pulps are extremely small and most of the substances are washed out from the fibers and found in the effluents. The appearance of these highly toxic substances caused much concern and extensive studies were made in Sweden by Christoffer Rappe (on careful methods of analysis) and Knut Kringstad (on the formation during the bleaching processes).

It is now established that chlorinated dioxins are formed during burning of chlorine-containing organic material. The cellulose industry is responsible for only a small part. Even so, the dioxin formation in the cellulose processes ought to be eliminated. Knut Kringstad and his research group at the Swedish Forest Products Research Institute (STFI) in Stockholm have recently shown that the dioxin formation can be reduced and practically eliminated. In genera, bleaching with oxygen and chlorine dioxide instead of chlorine gas reduces the dioxin formation. Chlorine dioxide causes no dioxin formation at all. It is possible now to apply a pulping and bleaching process, which according to Knut Kringstad gives a paper pulp with satisfactory brightness and practically no dioxin formation in the process.

A related problem is the chlorinated organic compounds which are residues from bleaching of lignin in the pulps and are polluting effluents from the mills. They are only slowly decomposed biologically in nature. The environmental protection agencies in Sweden have ruled that the amounts of total organic chlorine compounds (TOCl) must be decreased in the pulp mill effluents from the present level of 3-5 kg TOCl to 1.5 - 2 kg per ton pulp within four years. This means also a reduction in the amounts of chlorinated dioxins formed in the pulp mills.

There is a strong drive in Sweden to use more unbleached paper products. This will give higher yield of paper from the wood raw materials, it will save energy and chemicals and also give less polluting effluents and lower the amounts of toxic substances formed.

270

This review is based on the following references:

1. Original papers and reviews in Svensk Papperstidning (Swedish Paper Journal) for 1987 and 1988.

2. Annual Reports on Advances in Science and Technology for 1987 and 1988 by the Royal Swedish Academy of Engineering Sciences.

271

INDEX OF CONTRIBUTORS